Springer Biographies

The books published in the Springer Biographies tell of the life and work of scholars, innovators, and pioneers in all fields of learning and throughout the ages. Prominent scientists and philosophers will feature, but so too will lesser known personalities whose significant contributions deserve greater recognition and whose remarkable life stories will stir and motivate readers. Authored by historians and other academic writers, the volumes describe and analyse the main achievements of their subjects in manner accessible to nonspecialists, interweaving these with salient aspects of the protagonists' personal lives. Autobiographies and memoirs also fall into the scope of the series.

Springer Biographies

The books published in the Springer Biographies tell of the life and work of scholars, innovators, and pioneers in all fields of learning and throughout the ages. Prominent scientists and philosophers will feature, but so too will lesser known personalities whose significant contributions deserve greater recognition and whose remarkable life stories will stir and motivate readers. Authored by historians and other academic writers, the volumes describe and analyse the main achievements of their subjects in manner accessible to nonspecialists, interweaving these with salient aspects of the protagonists' personal lives. Autobiographies and memoirs also fall into the scope of the series.

Lawrence Mysak

Adventures in Climate Science, Ocean Waves, and the Flute

A Memoir

Springer

Lawrence Mysak
Department of Atmospheric and Oceanic
Sciences
McGill University
Montreal, QC, Canada

ISSN 2365-0613 ISSN 2365-0621 (electronic)
Springer Biographies
ISBN 978-3-032-19847-1 ISBN 978-3-032-19848-8 (eBook)
https://doi.org/10.1007/978-3-032-19848-8

Cover credit: Owen Egan/Joni Dufour

This Springer imprint is published by the registered company Springer Nature Switzerland AG
The registered company address is: Gewerbestrasse 11, 6330 Cham, Switzerland

In Memorium
Mary Mysak (1935–2011)

To
Paul and Claire
and
Janet

Foreword

It is a pleasure and an honor to contribute this foreword to Lawrence Mysak's *Adventures in Climate Science, Ocean Waves, and the Flute: A Memoir.*

I first met Lawrence Mysak in 1989, shortly after having joined McGill as a young Associate Professor in the Department of Mathematics and Statistics. Even though I was a junior faculty member from a different department, he was, in keeping with his character, very welcoming and supportive. He showed genuine interest in my research and was also keen to find out about my non-academic pursuits, so that we quickly found out that we shared a deep love of music. Thus started a wonderful friendship that has extended over three decades and that has meant a lot to me and my family.

I was very happy when Lawrence Mysak told me that he was working on a memoir, knowing that he was an excellent writer who would have a lot to offer to the reader. The wonderful book that resulted from this endeavor has in fact exceeded my highest expectations.

Indeed, Lawrence Mysak's memoir is remarkably different in both spirit and structure from the traditional autobiography one expects to see coming from a world-renowned scientist. Rather than providing a linear narrative anchored around a distinguished scientific career, he is inviting us to undertake a different and much more interesting kind of journey, one built on a beautiful mosaic of thirty-three exquisitely crafted short chapters, each capturing a significant episode of his academic path or a particularly meaningful moment of his rich personal life.

These chapters include moving accounts of his experiences growing up in a loving and supportive Ukrainian family in the Canadian prairies and vivid descriptions of some of the remarkable events and unexpected circumstances that helped shape his highly distinguished career as a researcher and academic mentor. This is a career that he started very precociously as a 21-year-old lecturer at the University of Alberta, after which came research on black holes at the University of Adelaide and a Ph.D. in applied mathematics completed in record time at Harvard. We then see how his research interests gradually evolved toward Ocean Waves and Climate Science, domains in which he quickly became a world-renowned luminary. The memoir also features brief but very meaningful glimpses into the rich musical component of his life as a gifted flautist, and, most importantly, contains vivid recollections of several particularly touching episodes of his family life.

The experiences shared by Lawrence Mysak in this memoir are conveyed with great evocative power and with the support of many wonderful photographs, inviting the reader to put his sensitivity and imagination to work to weave the fabric of a unique and fascinating life journey. They are also described with exemplary honesty by recounting how the challenges and disappointments that are inherent to even the most successful careers and the most fulfilling lives had to be overcome.

After having read this memoir, one cannot fail to be struck by the exceptional combination of qualities and values that have permeated Lawrence Mysak's life.

First come to mind his profound and unfailing love for his family and his deep attachment to his Ukrainian roots. No less striking are his boundless energy and drive and the crucial role played by his immense intellectual curiosity in his extraordinary academic successes. One also shares in his joy at experiencing the beauty and diversity of the world in all its forms through his many travels and experiences, and one is moved by his exemplary generosity in giving back as a mentor, benefactor and leader of major national and international scientific organizations. Finally, one is impressed throughout this memoir by his total absence of conceit, a precious quality that is typically the hallmark of the greatest scholars and scientists.

I am confident that the reader will derive tremendous enjoyment from reading Lawrence Mysak's memoir. It perfectly reflects the warmth and radiance of the author's personality, while giving glimpses into the mind and heart of an exceptional scientist and human being.

Niky Kamran, FRSC
Membre de l'Académie Royale de
Belgique
, Distinguished James McGill
Professor of Mathematics
McGill University
Montreal, Canada

Preface

Step into my home study and you'll find shelves overflowing with autobiographies and biographies—testament to a lifelong fascination with the stories behind great minds. Since my teenage years, I've drawn inspiration from the lives and work of renowned mathematicians, scientists, musicians, artists, philosophers, environmentalists, athletes, and influential leaders. When I retired from McGill in 2010, I started thinking about sharing my own story: a journey from the windswept Canadian prairies to a career as a mathematician, oceanographer, climate scientist, and professor at two leading Canadian universities. More than anything, I wanted my children and grandchildren to glimpse the experiences that shaped me, especially those rooted in my Ukrainian heritage in western Canada.

About a decade ago, I thought it was time to get to work. I didn't want to write a conventional, beginning-to-end autobiography, but something more personal. Janet, my wife, encouraged me to take a four-month memoir-writing course at Montreal's Thomas More Institute in 2018. That course changed everything and set me on my way. The instructors, Pauline Beauchamp and Karen Nesbitt, were inspiring and supportive. Each week they gave us a theme—a chance encounter, a life-changing experience, a favorite trip, a teacher who had left their mark—and challenged us to write an 800-word story. Those assignments opened the door for me, and I posted five of my early efforts on my McGill website. https://www.mcgill.ca/meteo/facultystaff/mysak/.

When the course ended, Pauline said to me, "You have an impressive CV and certainly a lot of stories to tell." That encouragement stuck. Over the next couple of years, I did more memoir writing between other adventures: donating my papers to the McGill Archives (Accession number 2021-0014; Fonds number MG 4333), biking through Europe, and celebrating my 80th birthday in January 2020 with Janet and kiwi friends in New Zealand.

Then COVID-19 hit. While many people used the lockdown to start or continue their writing, I didn't. Janet and I spent our time at the country cottage, staying active outdoors. It wasn't until the fall of 2021 that I felt ready to write again. This time, I added photos and drew on journals I had kept for more than 50 years—records of travels, major events, and books that had shaped me. Those journals brought the memories back to life.

The stories grew longer, richer, and more varied: family, mentors, university studies, teaching and science, music, nature, and travel. Along the way, Janet, my children Paul and Claire, and a circle of friends offered thoughtful comments (see Acknowledgements). Then, in 2022, Josef Schmidt—a retired McGill colleague and German literature scholar—generously stepped in as my editor-at-large. I'll admit, it wasn't always easy to hear his critiques or to revise a story several times. But in the end, the writing was improved, and the stories became closer to what I hoped to capture.

From 2022 to 2024, I shared selections of my memoirs with graduate students at McGill, Concordia, and the University of Alberta, and with seniors in Montreal residences. The feedback I received from family, friends and colleagues convinced me to think bigger. With the help of my longtime colleague Robert Doe at Springer Nature, I submitted a proposal to the Springer Biographies series. When Robert wrote to me in late 2024 to say that the reviewers had given their approval for publication with the series, I was thrilled.

This book is the result, after eight years of thinking and writing. My hope is to inspire *both* young people to consider studies in science and established scientists not to be afraid of a mid-career change. Above all, I'd like to give general readers a sense of what makes up the life of one scientist—its challenges, its surprises, and its rewards. I've tried to write these stories in a straightforward, non-technical way so that they can be enjoyed by most readers. I'm sorry to disappoint those seeking my favorite set of governing differential equations for the evolution of the climate system.

At heart, this book is about balance—between work and life, ambition and joy. As Roger Federer reminded graduates in his 2024 Dartmouth College convocation speech, "There is more to life than the tennis court." For me, that advice has played out in balancing professional work in science with my love of music, travel, nature, family, and friendships around the world. This has led to a deeply rewarding life.

In writing these stories I have, at age 85, become even more aware that my generation has had enormous luck: we have been able to explore many opportunities, enjoy new discoveries, and benefit from increasing economic well-being. Unfortunately, we are also leaving the world with many unsolved problems and unsettling situations. I'm not sure that future generations will be as blessed as mine has been, but I certainly hope that they will find better ways to live in harmony with the earth and with each other.

Montreal, Canada | Lawrence Mysak

Acknowledgements Over the decade during which this book was conceived and written, many have given me ideas, inspiration, support and encouragement. Early in our relationship, my wife, Janet Boeckh, asked me to write about what it was like growing up in a Ukrainian community on the Canadian prairies. Early one morning, I duly drafted a five-page outline for a traditional autobiography, which I then found had little appeal. However, I'm very grateful for Janet's next suggestion—to take a memoir-writing course at Montreal's Thomas More Institute in 2018. This helped me take off. Pauline Beauchamp and Karen Nesbitt led inspiring classes and gave helpful feedback on how to make my stories come alive and bring readers to the front row. I especially appreciated Pauline's encouragement to continue writing episodic memoirs after the course finished. I did this sporadically for the next few years. I am indebted to Janet, my first reader, for going over each draft and providing excellent suggestions for improvement.

A chance meeting of retired German studies professor Josef Schmidt at a 'round table' lunch at McGill's faculty club in February 2022 was another turning point in this venture. Josef offered to be my literary critic for any future memoirs that I might write. He kindly did this for the next three and half years, and the book is much better because of this valued input, which I absorbed while we sipped cappuccinos at the Mouton Noir café in our neighborhood.

Once I had written about 35 stories on a wide range of topics, i.e., family, science and academic life, music, nature and travel, and other loves and adventures, I started thinking about publishing some of them. But which stories should I choose? I sought the advice of Bob Stanley, the wise and longtime treasurer of the McGill University Retirees Association (Association des retraité(e)s de l'Université McGill). After reviewing the list of memoirs and reading those on my website, he recommended that a book focusing on climate science and music would have wide appeal. Thanks, Bob, for this timely and helpful suggestion.

I am very grateful to Dr. Robert Doe, Editorial Director, Applied Sciences, at Springer Nature for encouraging me to submit a proposal to the Springer Biographies series. Happily, my proposal was accepted in November 2024. I also appreciate the assistance of Springer Nature Production Editor Arumugam Deivasigamani and his team over the past year. Thanks also to Satish Ambikanithi and his team at Straive for help with the eProofing.

Throughout the writing of this book, I have received technical, artistic and related help from several people. I especially want to thank Ambrish Raghoonundun, a systems administrator from McGill's Faculty of Science IST (Information Systems and Technology) department. He graciously helped me set up the Word Template for the book, and he was always available for assistance when computer problems arose. Ambrish was particularly helpful with the submission of all the files for the book to the publisher Springer Nature and the eProofing of the the page proofs. Other IST staff who have helped me over the years include Brigitte Dionne, Calin Giurgíu, Michael Havas and Joseph Vacirca. For the insertion of the photographs and figures, I thank Nevein Gamal, Olivia Marino, Claire Mysak, and Joseph Vacirca. Unless otherwise specified, the photographs from 1972 to 2011 were taken by my late wife Mary Mysak. After 2012, most of the photographs were taken either by Janet Boeckh or me.

I am grateful to Djordje Romanic and Hannah Blumenthal for drawing Figs. 17.1 and 21.1 respectively. Thanks also to Hannah Blumenthal for getting permission to use copyright material and for helping me create a list of alternative texts for the figures and photographs.

It is a great pleasure to thank many friends, colleagues and family members for providing feedback and reviews for one or more chapters in the book. They have also helped to eliminate typos, inconsistencies and grammatical errors. While they are listed in the acknowledgements of the individual chapters, I list them all here in alphabetical order:

Nicholas Acheson, Thomas Beer, Claudia Bierman, Joyce Blond, George Bluman, Karen Boeckh, Gail Chmura, Nikolay Damyanov, Agatha de Boer, Jacques Derome, Roussos Dimitrakopoulos, Mary Feher, Michael Flanders, Doreen Friedman, Leon Glass, Chris Green, William Hsieh, Moto Ikeda, Shiro Imawaki, Alexandra Jahn, Linda Johnston, Niky Kamran, Lanny Levine, Charles Lin, Davinder Manak, Juliana Marson, Damon Matthews, Robert McKellar, Amal Mikdame, Timothy Moore, Claire Mysak, Marlene Mysak, Paul Mysak, Gerbern Oegema, Daphne Osoba, Wayne Pollard, Milica Radisic, Paola Rizzoli, Djordje Romanic, Nigel Roulet, Gavin Schmidt, John Smol, Neil Stewart, Thomas Stocker, Gordon Swaters, Richard Thomson, Bruno Tremblay, Hank van Harren, Silvia Venegas, Zhaomin Wang, Dan Wayner, Andrew Weaver, Andrew Willmott, and Ireneus Zuk.

Although I retired from teaching in 2010, I have been lucky to have retained an office at McGill where I have done most of the writing for this book. I thank John Gyakum and Daniel Kirshbaum, former chairs of Atmospheric and Oceanic Sciences, and Andreas Zuend, the current chair, for allowing me to have this privilege. Andreas has also provided me with office supplies and computer accessories. Over the past two decades I have also benefited from the support and encouragement from past deans of Science, Martin Grant and Bruce Lennox. Thank you, Martin and Bruce.

I was thrilled when my longtime friend and colleague Niky Kamran agreed to write a foreword to this book, providing I would first give him a draft of all the chapters. I am most indebted to Niky for his warm thoughts and kind words in the foreword.

Finally, I wish to acknowledge the support and encouragement from family members while writing this book. I thank Janet and my children Paul and Claire for patiently listening to weekly reports on my progress with the book. I also thank my sister Helen Osoba, niece Daphne Osoba, nephew Greg Osoba, cousins Marianne Carr, Joanne Mysak and Marlene Mysak, and Janet's sons Ian Boeckh and Rob Boeckh for their abiding support.

Competing Interests The author has no competing interests to declare that are relevant to the content of this manuscript.

Contents

Acronyms

AE	Academia Europaea
AES	Atmospheric Environment Service (Canada)
AGU	American Geophysical Union
AMS	American Meteorological Society
AOS	Atmospheric and Oceanic Sciences (Department, McGill University)
C^2GCR	Centre for Climate and Global Change Research (McGill University)
CCSM3	Community Climate System Model Version 3 (NCAR)
CM	Member of the Order of Canada
CMOS	Canadian Meteorological and Oceanographic Society
CSIRO	Commonwealth Scientific and Industrial Research Organization (Australia)
DAMTP	Department of Applied Mathematics and Theoretical Physics (University of Cambridge, England)
EC	Environment Canada
EGU	European Geosciences Union
EMIC	Earth system Model of Intermediate Complexity
ENSO	El Niño-Southern Oscillation
ETHZ	Eidgenössische Technische Hochschule Zürich
FRS	Fellow of the Royal Society (London)
FRSC	Fellow of the Royal Society of Canada
GFD	Geophysical Fluid Dynamics
GISS	Goddard Institute for Space Sciences (NASA)
IACS	International Association for Cyrospheric Sciences (IUGG)
IAHS	International Association of Hydrological Sciences (IUGG)
IAMAS	International Association for Meteorology and Atmospheric Sciences (IUGG)
IAPSO	International Association for the Physical Sciences of the Oceans (IUGG)
IGBP	International Geosphere Biosphere Program
IPCC	Intergovernmental Panel on Climate Change
ISC	International Science Council

IUGG	International Union of Geodesy and Geophysics
IUTAM	International Union of Theoretical and Applied Mechanics
JHU	Johns Hopkins University (Baltimore, MD)
MAE	Member of Academia Europaea
MIT	Massachusetts Institute of Technology (Cambridge, MA)
MOIST	Meteorological and Oceanographic Influences on Sockeye Tracks (UBC and Fisheries and Oceans Canada Research Project)
MPM	McGill Paleoclimate Model
MURA	McGill University Retirees Association
NAO	North Atlantic Oscillation
NAS	National Academy of Sciences (US)
NASA	National Aeronautics and Space Administration (US)
NATO	North Atlantic Treaty Organization
NCAR	National Center for Atmospheric Research (Boulder, CO)
NGO	Nongovernmental Organization
NRC	National Research Council (Canada)
NSERC	Natural Sciences and Engineering Research Council (Canada)
OC	Officer of the Order of Canada
ONR	Office of Naval Research (US)
OSS	Outdoor Skating Season
PNA	Pacific North America (teleconnection pattern in the atmosphere)
Q&A	Question and Answer
RA	Research Assistant
RS	Royal Society (London)
RSC	Royal Society of Canada
SCOR	Scientific Committee on Oceanic Research
SO	Southern Oscillation
THC	Thermohaline Circulation
TMI	Thomas More Institute (Montreal)
U of A	University of Alberta (Edmonton)
UBC	University of British Columbia (Vancouver)
UQAM	Université du Québec à Montréal
UVic	University of Victoria (British Columbia)
WCRP	World Climate Research Program
WMO	World Meteorological Organization

Part I
Growing Up in a Ukrainian Canadian Prairie Family

Chapter 1
Mom and Dad

Abstract My parents were born to immigrant families who came to Canada from western Ukraine before World War I to be pioneer farmers in the Canadian prairies. Mom and Dad met in Saskatoon, Saskatchewan when studying to be teachers. After they spent several years teaching in one-room schools in rural Saskatchewan, my father studied agriculture during the 1930s. During World War II, he served as a recruitment officer in western Canada for the Canadian Army.

After the war, when my sister Helen and I were 14 and 6, we settled in Edmonton, Alberta where there was a large Ukrainian community. Dad was at first employed by Agriculture Canada there, but he switched to selling insurance and real estate before rounding out his career as a high school business teacher. Mom returned to teaching when I was in high school. Mom and Dad remained in Edmonton after retirement in 1972. After my mother's death in 1978, Dad moved to Vancouver, British Columbia to be near Helen and me; he died there in 2007, at the age of 100.

In July 1972, I was waiting for my parents at busy Heathrow airport in London, UK. They had a long flight from Edmonton, Alberta, but looked happy exiting from customs. We were about to begin the holiday of their lifetime, spending 10 days touring England and Scotland and then 10 days in continental Europe.

Now I was thrilled that I could share with them many of the sensory delights that I had experienced during my sabbatical year in Cambridge. Mom and Dad had both grown up on farms and had lived through two world wars and a depression. They had always worked hard and never had an easy life. This was their first trip to Europe.

1.1 My Mother

Anastasia (Nettie) Trojan was born the youngest of three sisters on January 7, 1907, in Gimli, Manitoba. Her parents, George Trojan and Ewdochia (Dora) Sherbluk, had come separately with their parents to Canada from western Ukraine in 1899. They married a year later. After 10 years in Gimli, her family of five moved, with two cows

L. Mysak, *Adventures in Climate Science, Ocean Waves, and the Flute*, Springer Biographies, https://doi.org/10.1007/978-3-032-19848-8_1

Photo 1.1 Petra Mohyla institute in Saskatoon, Saskatchewan, in 1925

and an ox in tow, to central Saskatchewan, where the farming conditions were much better. It took 3 years to build their house and barn on new farmland about 80 km east of Saskatoon. After the sisters Mary, Ann and Nettie, the Trojans had two sons, Alexander and Peter. Mary and Alexander stayed in Saskatchewan, Ann moved to Winnipeg, Manitoba, and Peter settled in Vancouver, British Columbia.

Mom started her education in a one-room schoolhouse founded by her father near the farm and moved to the railroad town of Colonsay to complete grade 8 at the age of 15. In Saskatoon she took grades 9–12 at Nutana Collegiate and then completed a 1-year teacher training course at the Normal School in 1926. While studying in Saskatoon, Mom stayed in a residence for young adults of Ukrainian ancestry known as the Petra Mohyla Institute (Photo 1.1). I believe this is where Mom (Photo 1.2) met Dad when she was 17.

1.2 My Father

Stephen Mysak was born on December 24, 1906, in western Ukraine in the village of Kluventsi, near Ternopil. He and his mother, Elizabeth Mysak, emigrated to Canada in April 1912 to join his father, Ivan Mysak, who came in 1908 to work and establish a farm north of Regina. Dad and his mother had quite an adventure while sailing to Canada. Their ship, the Carpathia, was near the ill-fated Titanic and rescued 800 passengers. Dad remembered this tragic event all his life.

Photo 1.2 Mother (on the right) together with her oldest sister Mary (on the left) and younger brother Peter (in the middle) in 1924. Their mother, Ewdochia (Dora) Trojan (in black), was widely known in the countryside for her outstanding work as a midwife

Later Ivan and Elizabeth Mysak had four more children—Walter, Olga, Arthur and Orest. My favorite uncle was Orest, who was 18 years younger than Dad. Arthur inherited the family farm, and Walter and Orest, like Dad, studied agriculture at university and remained in Saskatchewan. Olga served in the war and settled in Thunder Bay, Ontario.

As a teenager (Photo 1.3), Dad left the farm for high school in Saskatoon, where he stayed at the Mohyla Institute. Like Mom, he also completed the 1-year teacher training course at the Normal School in 1926 when he was 19.

1.3 They Meet

I suspect Dad and Mom took a liking to each other when they were 17 or 18 and boarding in the Mohyla Institute. In the late 1920s they both taught in one-room schoolhouses in northern Saskatchewan, but they never worked in the same area. I often wonder how they kept a romance going—did they travel by horse and buggy to

Photo 1.3 Father at age 18 in 1925 when living in the Petra Mohyla Institute in Saskatoon

meet each other at some mid-point? Or did Dad borrow a car to make visits to Mom? I know they were great letter-writers. Sadly, Mom threw out their letters many years later when house clearing.

Mom and Dad married in 1930 when they were both 23 (Photo 1.4). Dad looked very serious that day.

In contrast to Mom and Dad, Mom's parents (Photo 1.5) married as teenagers in 1900. Dad's parents married in 1905 and celebrated their 50th wedding anniversary in 1955 (Photo 1.6).

1.4 The 1930s

The 1930s were a tough time for my parents. When the depression came in the early 1930s, teachers' salaries were cut and life was a challenge with a young daughter, my sister Helen, born in January 1932. As a married woman with a child, Mom could no longer teach. With the help of the Reserve Officers' Training Corps (ROTC) program, Dad was able to study farming at the University of Saskatchewan and obtain a BSc in Agriculture in 1936 (Photo 1.7). One fall, Dad could not pay his tuition fees in early September because the crops from the family farm had not yet been sold. His plea to the university bursar for an extension landed on deaf ears. Undeterred, Dad went to the president's office to seek help. The president took Dad's hand and walked him to the bursar's office and said, "Please make out a promissory note for Stephen Mysak which allows him to pay his tuition in a month's time." Hence Dad started

Photo 1.4 Mom and Dad's wedding portrait taken on July 15, 1930

his classes immediately that term. I wonder how many university presidents would now take a similar step.

Dad was a good student and after graduation was encouraged by his professors to conduct summer research projects on land management. This led to an MSc in Agricultural Economics in 1939, and his job prospects looked excellent. Unfortunately, his plans changed when World War II began in the fall of 1939 and Mom was pregnant with me.

The Trojan family suffered the loss of two of their children during the 1930s. Mary died after a botched operation in a country hospital, and Alexander died of an infection that developed while working on a cadaver in medical school at the University of Saskatchewan. Mom often talked about her two siblings who died so young. To honor her late brother, Mom and Dad chose Alexander as my middle name.

Photo 1.5 Mom's parents Ewdochia (Dora) and George Trojan in 1928. Dora (1881–1948) died when she was 66. George (1982–1972), however, lived to 90. I had the impression that they were both rather serious and strict with their children

Photo 1.6 Dad's parents in 1955 at their 50th wedding anniversary. Ivan Mysak (1882–1958) lived to 76 and predeceased his wife, Elizabeth Mysak (1888–1963) by 5 years. Elizabeth, who lived to 75, had quite a sense of humor that she passed on to Dad. Years ago, when I was picking Saskatoon berries with her on the family farm, I noticed with horror tiny larvae on a few of the berries. "Don't worry," she said, "these are good for your digestion"

Photo 1.7 Dad at his graduation from the University of Saskatchewan in 1936. He was 29

1.5 The War Years, 1939–1945

Dad served as a Captain in the Canadian Army for the entire war. He was nearly sent overseas—twice he went to Halifax to board ship—but each time he returned to serve as a recruitment officer, traveling widely across western Canada. After I was born in January 1940, Mom, Helen, and I spent 2 years in Saskatoon; I spoke Ukrainian as my first language. As rental accommodation was scarce in Saskatoon, we had to move out of the city to the town of Colonsay, where we were lucky to find a simple but rundown house to rent for the next 2 years. There was running water but no indoor toilet. Not much fun during winter. On the positive side, we were close to Mom's family farm where food was plentiful.

While living in Colonsay, we spent the summers in Chilliwack, BC, where Dad was training with the army. We also had lovely holidays in Cultus Lake nearby (Photo 1.8). Helen loved to read, and she would climb into a tree with a book and ignore my plea, "Helen, I want to play with you. Where are you?"

The train rides between Chilliwack and Colonsay seemed endless, especially on the prairies. To pass the time, Mom asked us to count telephone poles, look for interesting birds, and check out the farmyards that we passed. But she was most embarrassed 1 day when I said to her in my loud voice, "Look at that shack, mummy, it's just like our house." This is a story I heard repeatedly from Mom.

The last 2 years of the war were spent in Winnipeg as Dad was doing further training at Camp Shilo to the south. Dad was now spending more time with the family, and Helen was grateful that she could ask both Dad and Mom for help with

Photo 1.8 Helen (10½), me (2½) and Dad (35) in summer army uniform at Cultus Lake, BC

schoolwork. Before I started school at 5½, I was very restless and always wanted to be doing things, driving my parents crazy.

"I don't know what to do!" I said to Dad one day. "Jump up and down," he replied.

When I was about five, I became interested in model-building. I tried to make a balsa wood airplane from a model kit, but I gave up because I could not read the instructions in the box. Then Mom said, "Why don't you try building something with toothpicks and glue?" And I did, making log cabins and lookout towers. With this new hobby, I became less of a pain for Mom and Dad.

When the war ended, we had a family portrait taken (Photo 1.9). Mom and Dad were now smiling, and Helen was developing into a beautiful young lady. I don't look happy in this picture because I had to be still for what seemed like ages while the photographer arranged our seating.

Photo 1.9 Mom and Dad at age 38 with Helen (13½) and me (5½) in 1945

1.6 The Move to Edmonton

We moved to Edmonton in 1946 where Dad did farm surveys for Agriculture Canada. He had a government car, and we took road trips to Saskatchewan, where much of his work was focused. Mom, Helen, and I thus often spent summers at the Trojan family farm. I am sure the reports about his surveys were excellent, but he often said most of them gathered dust in Ottawa.

In Edmonton, Mom taught Ukrainian on Saturdays at "Ridney Shkola," a school for Ukrainian language and culture. Helen and I were not keen students, and I dropped out after a few years, much to Mom's disappointment. While I learned to converse in Ukrainian, I never became fluent in the language. However, even this rudimentary knowledge of Ukrainian was most helpful during my visits to Ukraine in 1972, 1993, and 2002.

On Sundays we went to endlessly long Ukrainian Orthodox church services and attended Sunday school classes. Often my parents invited an out-of-town guest at the service for a roast chicken lunch. We had lively conversations which I found more interesting than the service.

1.7 Music and Food

As in many Ukrainian families, music played a big part of our lives. Mom and Dad loved to dance, and Dad played the mandolin and had a fine tenor voice. He also conducted school choirs when he was teaching. We listened to classical music on the radio, and later we attended symphony concerts.

Helen had a beautiful soprano voice and sang in the church choir and the Edmonton Youth Choir. She also took private voice lessons, and Mom and Dad were so proud when at age 20 she won a scholarship to study music and singing at the Banff School of Fine Arts (now called the Banff Centre). After her marriage to David Osoba in 1953, she sang in top-level amateur choirs in Vancouver and Toronto. Helen later trained to become a music therapist; she worked in senior residences and had her own private practice.

In 1948, Mom bought a used piano with money earned teaching Ukrainian, and I started piano lessons. I went as far as grade 7 with the Toronto Conservatory of Music but switched to the flute when I was 12. At that time, I had to decide between conflicting evening band and hockey practices. Dad helped me choose one of these. "I doubt if you will be playing hockey at 50, but you might still be playing the flute," he said. Dad was so right! (See Chap. 28).

Helen also learned to play the piano, and later she and I enjoyed playing simple flute-piano duets. Mom and Dad encouraged me to continue with flute lessons while playing in the Edmonton School Boys Band, and in 1958 at age 18, I passed the exam for the Performance Diploma at the University of Alberta. Over the years, I have played flute in four amateur orchestras, in Edmonton, Cambridge (England), New Westminster, and Montreal.

Food has always been important to Ukrainians, and Mom was an excellent cook. We often had tasty cabbage rolls and perogies served with fried onions and sour cream. Her beet borscht, which also included a dozen other vegetables, was sublime. However, for a while she had to watch her diet and fed us wheatgerm wafers, vegetables steamed in parchment paper, and a lot of chicken and fish. Helen and I craved pies and ice cream, and after dinner we sometimes snuck out of the house for a milkshake at the corner ice cream parlor. Even today, my wife Janet claims that I still suffer from food deprivation.

1.8 The Late 1940s and 1950s

In 1948, Dad left his job with Agriculture Canada and tried his luck in business. He briefly owned a mushroom-growing company and later sold real estate and insurance. He was moderately successful at the latter but never made "big money" as a salesman. "I'm too honest to get rich working in the business world," he once told me. We could only afford to take modest holidays in the Canadian Rockies and in other resort areas (Photo 1.10), and we never stayed in fancy hotels. Mom usually had to cook our own meals—not any break for her! But Mom and Dad believed that a holiday was a time to see new parts of Canada as a family and to get to know the world outside the Ukrainian Canadian community.

In 1949, we moved to the south side of Edmonton, near the university. With the help of a veteran's grant, Dad purchased a lot in a new development and had a four-bedroom house built. One bedroom was intended for Mom's Dad, George Trojan, who was widowed the previous year in Saskatchewan. But instead, he remarried and

Photo 1.10 Mom and Dad on holidays in Prince Albert National Park, Saskatchewan, in 1953

continued working on the Trojan family farm. In any case, Dad ran out of money to finish the house, and we had to move to a rental property nearby. Eventually, after Grandfather Trojan's second wife died around 1970, he did come to live with Mom and Dad in Edmonton, where he died in 1972 at age 90.

The birth in February 1957 of Mom and Dad's first grandchild, Greg Osoba, the son of Helen and David, was an exciting time. I still remember the 1957 summer road trip to Vancouver with my parents and Uncle Orest and Aunt Mary to meet Greg. Helen and David later had a daughter, Daphne Osoba, born in January 1960. There were more happy trips by Mom and Dad to Vancouver to see the two grandchildren (Photos 1.11 and 1.12).

Photo 1.11 Greg (4) and Daphne (1) in their Vancouver home. Photograph courtesy of my nephew Greg Osoba

Photo 1.12 Greg (6) and Daphne (3) on holidays with their parents in Nimes, France, summer 1963. Photograph courtesy of my nephew Greg Osoba

After 10 years in business, Dad went back to university to upgrade his teaching qualifications. When I was doing my mathematics and physics degree at the University of Alberta, Dad was studying psychology and pedagogy in the Faculty of Education. In 1959, at the age of 52, he returned to teaching. It was around this time that the Canada Pension Plan was founded, and he wanted to be on the ground floor of this saving scheme for retirement. I admired him for taking up teaching again. It had been nearly 30 years since Dad had been in a classroom. For 13 years he taught business mathematics, bookkeeping, and economics at Ross Shepherd High School. The students were not the brightest, but he was highly regarded as a teacher. At retirement he was given a marble pen holder in appreciation for his work as Treasurer for the Edmonton Public School Board. I now have this souvenir in my McGill office.

In the early 1950s, Mom worked in the summer as an activities program leader in the public parks of Edmonton. With this experience and her Normal School training, Mom qualified for a full-time job as a kindergarten teacher at the army base in Edmonton. This work gave Mom much confidence and some financial independence, and together with Dad's good income as a high school teacher, they were finally able to buy a small semi-bungalow in 1961 after many years of renting. They were now in their mid-50 s. In 1963 they upgraded to an attractive three-room bungalow overlooking the North Saskatchewan River, thanks to an inheritance from Dad's mom, Elizabeth Mysak. But this inheritance came in a roundabout way. In her will, Dad was disinherited because he did not want her to come and live with us after her husband, Ivan Mysak, died in 1958. "She was too bossy," Dad said. When Dad's four siblings learned of this injustice, they divided the total inheritance into five equal parts and gave Dad his share.

Sometime in the late 1950s, Mom learned to drive the family car, which was a big step for her. She was then able to shop on her own and share the driving with Dad on their long road trips to Saskatoon, Vancouver, Winnipeg, and Boston.

1.9 Marriages and Breakups

In the 1960s, I was in graduate school, first at the University of Adelaide for an MSc in mathematics (Chap. 5) and then at Harvard University, for a PhD in applied mathematics and physical oceanography (Chap. 7). In 1965 I married Diana Shklanka, a brilliant English scholar and university lecturer. Our families knew each other from the Ukrainian community and thought that we were a good match. But before the marriage, we spent little time together and did not get to know each other well. After 2 years in Cambridge, Massachusetts, and 4 years in Vancouver, where I was now teaching mathematics at the University of British Columbia (Chap. 10), we broke up, which was not a surprise to my parents. In the summer of 1971, I went off by myself on sabbatical to the University of Cambridge in the UK (Chap. 11). While on sabbatical I met photographer Mary Burr, whom my parents met in the summer of 1972. They were skeptical that our relationship would last. "I can find another Ukrainian woman for you to marry," Mom said. She could be a bit bossy.

In 1969 Helen and David's marriage also broke up, which was upsetting for Mom and Dad. Dad encouraged Helen's long-term desire to get a university degree so that she could get a good job and become financially independent. Helen completed a BA in fine arts at York University in 1977 and worked there for a few years before moving to Vancouver, where she obtained a postgraduate diploma in music therapy at Capilano College in 1981. As mentioned earlier, she then worked for many years as a music therapist.

1.10 Later Years, 1972–1978: Happy and Sad Times

Mom and Dad retired from teaching in June 1972. They were both 65 and had pensions. Since I had been on sabbatical in Cambridge, England for nearly a year and had a car there, I encouraged them to join me for a road trip in the UK and in western Europe. They had never been on a holiday to Europe. They got a big kick out of feeding the pigeons in Trafalgar Square (Photo 1.13). After seeing London, Cambridge, Edinburgh, Manchester, and Newcastle (Mom's brother Peter Trojan had relatives by marriage whom we visited in the latter two cities), we ferried over to Belgium and continued our journey with stops in Brugge, Amsterdam, Heidelberg, Geneva, Paris, and Brussels. Mom got her wish to spend a night in a Swiss chalet in the Bernese Alps, and Dad got his fill of very fine wines in Paris.

Mom and Dad met Mary shortly after their arrival at Heathrow. She impressed my parents with a tender chicken-in-beer casserole dinner served in her London bedsit, and we enjoyed an evening concert featuring violinist Yehudi Menuhin. Mary, being English and not Ukrainian, knew it might be an uphill battle to be accepted by my parents. Dad was the first to see that Mary and I were very fond of each other, and that the relationship was serious. On the road trip with Mom and Dad, I persuaded Mom that Mary was for real, and I told them that I had invited her to join me in Vancouver in the fall.

Mary met us in Brussels at the end of the European road trip for a farewell dinner. That went very well, and I could see Mom now liked Mary. From Brussels Mom and Dad made their way home to Edmonton, while Mary and I continued by car and ferry to Helsinki, where we started a 2-week road trip to Russia and Ukraine. (Some of our experiences on this trip are described in Chap. 32).

After I returned to Vancouver in September, Mary joined me later in the fall and became a permanent Canadian resident in 1973. This allowed her to work part time, which give her some financial independence.

Mom and Dad did not come to Vancouver until our marriage in August 1974. Prior to the wedding, Mom sent me a recent portrait (Photo 1.14) which I have always liked.

With the help of a $10,000 loan from Mom and Dad in the fall of 1974, we bought a four-bedroom ocean-view home in Point Grey, near UBC. Shortly after moving in, we announced that Mary was expecting a child next June. Mom and Dad were thrilled. But this joy was short-lived. In early December, Mom was diagnosed with lymphoma. Mary and I went to Edmonton for Christmas and tried to convince Mom to have chemotherapy and radiation treatment. However, she refused these treatments; she would rather fight the cancer with special diets and quack medicines. It was a sad Christmas.

Photo 1.13 Mom, dad, and me feeding the pigeons in Trafalgar square, London in July 1972

In June 1975 Mary gave birth to a healthy son, Paul Alexander Mysak. Mom and Dad were so excited that they made a special road trip to Vancouver to see their newest grandchild (Photo 1.15). Although Mom was suffering from the effects of lymphoma, she disguised it well (Photo 1.16) and lived for another 2½ years. She died on February 22, 1978, 4 days after the birth of our daughter Claire Anastasia Mysak (Photo 1.17). Mom knew that Mary was going to have a girl, but she kept this as a secret until the birth. During 1975–1977, Mom's health had slowly declined, but she hardly ever complained. I suspected she was in pain much of the time. In the last few weeks of her life, Helen and a favorite niece Marlene Mysak came from Toronto and Vancouver respectively to provide support and comfort.

Photo 1.14 Nettie Mysak at 67 in summer 1974

Mom's funeral in Edmonton was held in St. John's Ukrainian Orthodox Cathedral where she had always been involved in church activities and was loved by many. There were over 200 at the funeral. I felt sadness in losing Mom and realized that there wouldn't be any more phone calls from her. And it really hit me that she would never see Paul and Claire in the years ahead.

As I was growing up, Dad, and especially Mom, did so much to support my activities outside school, whether it was sports (archery, hockey, badminton) or music (flute, piano) or hobbies (watercolor painting, woodworking, stamp collecting). Mom was sometimes over-controlling and was not pleased that I stopped going to church in my mid-20 s. "You can do what ever hobbies you like, but I want you to come to church with us," she often said. Despite my leaving the church, today I am sure she would be happy to know of my continuing contacts with the Mysak family and my interests in Ukrainian culture and history. She would have loved to have heard about my trips to Ukraine after its independence on August 24, 1991.

Mom was buried in St. Michael's cemetery in northeast Edmonton. About 15 years later, Ann Osoba, the mother-in-law of Helen, took me to her grave, where I spent a long time thinking about Mom. How I wished I could have told her about Claire's achievements in gymnastics and Paul's prowess in sports, and that they were both good students. And while she knew that I had been promoted to Full Professor at

Photo 1.15 Mom and dad with grandson Paul Mysak, when he was 8 weeks old, in August 1975

UBC at the age of 36, she would never know about my subsequent research and teaching achievements at UBC and McGill University.

1.11 Dad in Vancouver, 1979–2007

The year after Mom died, Dad sold the family home in Edmonton and moved to Burnaby, just east of Vancouver, to be near family. Helen and Daphne were also in the process of moving to Vancouver, and my uncle Peter Trojan was already living there. But Dad also wanted to give us breathing room. We would visit him on weekends for a delicious lunch that he prepared (yes, he was a good cook too), and he came to our Point Grey home near UBC to take Paul and Claire to the parks. Dad also had his own life. He went for walks, listened to music, studied current events and the stock markets, and exercised in the swimming pool. Dad also had a passion for bridge and joined several senior card clubs in Burnaby and nearby New Westminster. But he was lonely and thought he would like to get married again.

He invited Melany from Toronto to visit him in Vancouver. Melany, a widow he had known since high school in Saskatoon, accepted his proposal of marriage. She

Photo 1.16 Dad, mom, Mary, and me in August 1975 in front of our Vancouver house

Photo 1.17 Dad, Helen, and Claire Mysak, when she was 2 weeks old, in March 1978. In honor of Mom, Mary and I chose Anastasia as her middle name

moved to Burnaby, but the marriage only lasted 2 years. She missed her family in Toronto, and she got depressed by the lengthy rainy periods in Vancouver. I wondered about the relationship when I noticed that they had separate bedrooms. After they divorced, Dad had the habit of saying to Helen and me, "Now we all have something in common to talk about: a divorce."

In 1982–1983, I spent a sabbatical in Zürich (Chap. 14). Mary and I invited Dad to visit us in Switzerland and see a new part of Europe. He was then 76, and we were amazed to learn that before coming he studied German at night school. After his arrival in Switzerland, Dad took a bus tour to Vienna and Salzburg. He was in heaven listening to the music of Mozart and Haydn in the heartland of classical music. "Das ist Wunderbar," he said.

Sometime in the mid-1980s, Dad met Alice at a community dance club. She was married to a much older man who was the priest in the Ukrainian Orthodox church in Burnaby. Alice and the priest did not do much together, and the priest encouraged Dad to take Alice to the movies and dances. Dad was an excellent dancer, and the women on the dance floor enjoyed a swing with him. After a few years the priest died, and Dad and Alice were more open in their relationship. They never lived together, but they were an item for many years. He was very cheerful during this time. Sadly, Alice died in her early 90 s.

In 1986 when Dad was in his 80th year, my own family moved to Montreal. I was excited to be appointed to a new chair in climate research at McGill University (Chap. 15). Dad visited us every year as he wanted to follow Paul and Claire's progress. In addition, he was intrigued by my work at McGill, and he rejoiced in my success in the academic world. He hugely enjoyed lunches at the spectacular McGill Faculty Club [1], where one day he was thrilled to meet Principal David Johnston. I often went to the Ballroom of the Faculty Club for lunch because this is where I met many wonderful colleagues at McGill, most of whom were very welcoming to me as a newcomer on campus. More than once, Dad said, "I would have loved to have been a university professor." Shortly after the war, he received a government award to take his PhD in agricultural economics in the United States but decided against doing this with two growing children to support.

Over the next 15 years, Dad played many games of bridge in Montreal with my friends and family. The games with Sam Ducharme, Charles Lin, Dad, and myself were especially memorable. Sam was amazed how Dad could remember all the cards as they were played in a contract and then tell you how you could have gotten another trick! Paul had learned to play bridge with Mary and me in Vancouver as a youngster, and 15 years later, he and his friend Steve would challenge Dad and me to rubber bridge over a few beers and spicy chips in his apartment. Today, I still get a lot of pleasure playing bridge, both in person and online. I regularly participate in the bi-weekly games for retirees at the McGill Faculty Club, which also offers the chance to catch up with many former colleagues.

While in Montreal we frequently went back to Vancouver for visits and special birthdays. Dad always had a strong interest in the lives of family members. Since Dad's birthday was on Christmas eve, we only went to Vancouver for the big ones, like his 90th which was celebrated in 1996. On this occasion Mary took a "family portrait" of Dad with his children and four grandchildren (Photo 1.18a). This was followed by a special dinner at a local restaurant where Dad only had 9 candles to blow out (Photo 1.18b). The year 1996 was also a special one for me: with great pleasure I told Dad on his 90th birthday that I had just been named to the Order of Canada for my contributions to science.

When Mary completed her BA in Asian religions at McGill in 1999, Dad aged 92 came for the graduation ceremony and celebrations. Mary often admired a small oil painting of a house and garden that Mom and Dad had in the hallway of their

a

b

Photo 1.18 **a** Dad at age 90 with his children Lawrence (at 56) and Helen (at 64) in the front row, and grandchildren in the second row (L to R): Greg (39), Claire (18), Paul (21), and Daphne (36). December 24, 1996, in Vancouver. **b** Dad making a wish before blowing out the candles on his 90th birthday (December 24, 1996)

Edmonton home. Dad brought this picture to Burnaby after Mom died in 1978, and now he gave it to Mary as a graduation gift 21 years later. She was thrilled to have it in our home in Montreal West. Paul and his wife Allisha now have this picture hanging in their home in Prince George, British Columbia.

The next year, Dad also came to Montreal for Paul's and Claire's university graduations, at Concordia University (Montreal) and Bishop's University (Sherbrooke), respectively. He was very proud of their academic successes, and even though he was now having trouble walking, he wanted to be there with the family on these two happy occasions.

Dad never wanted to visit Ukraine when it was under communist rule, and when independence came in 1991, he felt too old to travel that far. However, he was a big help to me in 2002 when I was preparing scientific lectures for my upcoming lecture tour there in June. At our Montreal West home (Photo 1.19) he translated my opening remarks into Ukrainian and corrected me as I practiced this part of my lecture: not an easy task since I had not spoken much Ukrainian for several years. Mary and Helen joined me on this trip, and a highlight was meeting Rosalia, Dad's 80-year-old cousin, who lived in Ternopil, near Lviv, the largest city in western Ukraine. We also met her daughter and family. Dad asked me to give each member of the family $100 US, as a gift from himself; not a lot of money to him, but it meant much to the relatives, who were having a hard time adjusting to the new political and economic freedoms in Ukraine.

In 2003, at age 96, Dad moved into Chalmers Lodge, an assisted living residence in central Vancouver close to Helen. He was quite happy there and even managed to take the bus to Burnaby occasionally to play bridge with old card friends. By this time his eyesight was poor because of macular degeneration, and he had to give up reading which he used to love. In May 2004 at age 97 he came by himself to Montreal for the last time, and Mary and I took him to our new country cottage in the Eastern Townships (Chap. 29). "How is the farm? he used to say over the phone

Photo 1.19 Dad at 95 in Montreal West on our porch in May 2002

when making calls to him from the cottage. Then he would start to reminisce about life on the prairie farm where he grew up. Little did he know that 4 years later, our daughter Claire would live on a dairy farm near Sherbrooke, Quebec after marrying Dennis Taylor, whose family had lived on this property since 1796.

In the fall of 2005 Dad needed more help and moved into the Purdy Pavilion, a nursing facility associated with the UBC hospital. It was bright and cheerful, and Dad loved to watch the university students come and go from their classes. He was now confined to a wheelchair, but he still enjoyed visits from the family and having phone conversations. He still had an excellent memory and was of sound mind. Once a week a volunteer took him to the UBC village for a Starbucks coffee and visit.

In the fall of 2006, just before his 100th birthday, Mary and I established the "Stephen and Anastasia Graduate Fellowship in Atmospheric and Oceanic Sciences" at McGill University to honor my parents (Chap. 33) Dad was thrilled to learn about this award. As of 2025, over 10 graduate students have benefitted from this major award.

On Dad's 100th birthday, Helen arranged for a family dinner at the Sylvia Hotel, our favorite meeting place for special occasions. He enjoyed the outing (Photo 1.20) and received many good wishes and greetings (including those from the Prime Minister and Queen Elizabeth). But we sensed that his will to live much longer was ebbing away. At 2 am on Tuesday April 17, I got a call in Vienna from Mary who said Dad was quickly fading and that I must leave the European Geosciences Union conference for Vancouver as soon as possible. Mary and I managed to see Dad on Thursday April 19, and we were amazed to have a spirited chat about the conference in the city he remembered visiting 25 years ago. "And how was Vienna?" he asked. He died peacefully of natural causes early in the morning of Saturday April 21, 2007. Regretfully, he was alone when he died.

Photo 1.20 Dad (center) with Helen (on left) and me on his 100th birthday which was celebrated at the Sylvia Hotel in Vancouver on December 24, 2006

Photo 1.21 Helen (center) at 85 with her children Greg and Daphne on the right, and me (behind Helen) with my children Claire and Paul on the left. Photo taken on January 14, 2017, at the Sylvia Hotel, Vancouver. Photograph courtesy of Janet Boeckh, whom I married in 2015

Dad had left the church a few years ago and requested no funeral service. His body was cremated, and we buried his ashes in Vancouver's Jericho Park, in the back yard of our Montreal West home, and in the forest of our cottage in Quebec's Eastern Townships. I placed a few large stones over the burial site at the cottage, and I often sit on a nearby bench thinking about Dad and his long life. For so many years he offered great support in all my activities, and I often felt he lived my academic world vicariously. One of the lessons I learned best from him was "never try to carry the world on your shoulders." By this I think he meant, "do the best you can to make the world a better place, but also accept your limitations and shortcomings."

1.12 Coda

Mary died suddenly of a hemorrhagic stroke in December 2011, only 4 years after Dad's passing. In the following year I was lucky to meet my future wife Janet Boeckh, a former high school teacher and career counselor (Chap. 21). We continue to have small family reunions and celebrations, like the one in January 2017, a decade after Dad died. This time we gathered in the Sylvia Hotel for Helen's 85th birthday (Photo 1.21). Today, I am happy that my children Paul and Claire have gotten to know Helen's children Greg and Daphne and that they will organize their own reunions in future.

Acknowledgments I thank my wife Janet Boeckh for her comments on the first draft of this chapter, Claire Mysak and Olivia Marino for photographic assistance, and Josef Schmidt for his literary critique.

Reference

1. Stewart NF (2024) From the Baumgarten house to the McGill faculty club. Marking the centenary of the faculty club 1924–2024. Accent Impression, Montreal QC

Chapter 2
Early Years: What Do I Want to be?

Abstract After spending the World War II years in several cities and small towns in western Canada, I moved with my parents and sister to Edmonton, Alberta in 1946 to complete my schooling and undergraduate university studies. This chapter is divided into four parts, my childhood years (1940–1951), junior high school years (1951–1954), high school years (1954–1957), and my 4 years at the University of Alberta (U of A) (1957–1961).

Finding my way to a career in mathematics and physics was not easy since I had many interests and hobbies when I was growing up, including music, boy scouts, woodworking, Ukrainian handicrafts, archery, hockey, baseball, and dancing. Under pressure from my parents to choose a "profession," in my last year of high school I decided that I would go to university to become a dentist. However, after 1 year in an unpleasant zoology lab together with an enjoyable calculus class, I switched to the second year of the honors applied mathematics program at the U of A. Throughout high school and university, I also studied music theory and flute performance with the then Western Board of Music, affiliated with the U of A. In the late 1950s, I also began to think seriously about a music career, and in pursuit of this possibility, I went off to study advanced flute performance at Aspen Music School (Colorado) in the summer of 1958 (Chap. 26). However, the competition was fierce, which convinced me that a career in university teaching was more to my liking. So off I went to Adelaide, Australia in January 1962 to begin graduate studies in mathematics under the sponsorship of a Rotary Fellowship for International Understanding.

2.1 Childhood: 1940–1951

Because Dad served as an Army Captain during World War II, we were always on the move. I lived with my mother and sister in five different places in western Canada during the first 6 years of my life. We usually moved to be close to where my dad was stationed. We spent my first 2 years in Saskatoon (Photo 2.1) and finally settled in Edmonton in 1946.

L. Mysak, *Adventures in Climate Science, Ocean Waves, and the Flute*, Springer Biographies, https://doi.org/10.1007/978-3-032-19848-8_2

Photo 2.1 The author sitting on the railing of the front steps of our house in Saskatoon, Saskatchewan in 1941 (age 1½). Mom used to say, "I had boundless energy and could be quite a handful"

In Edmonton, we first lived in old rental houses before moving when I was nine to be near the University of Alberta and better schools. As I was a good student in my first few years of school, my parents were sure that I would go on to university one day. I kept busy with lots of hobbies (woodworking, model building (Photo 2.2), chemistry experiments) and piano lessons, which I started at age eight.

Edmonton had a large Ukrainian community, and my parents were active in helping to build St. John's Ukrainian Orthodox Cathedral. They wanted me to improve my Ukrainian (my first spoken language) by going to Saturday morning language classes. After a few years I balked at this as I found learning English and Ukrainian to be confusing. For example, the Ukrainian written letter "m" is

Photo 2.2 Me with the boat I built that wouldn't float in 1948. As a youngster, I loved the reedy smell of lake water. At age 15, I finally built a non-leaking 10-foot rowboat which my parents and I often used for fishing in prairie lakes

pronounced like the English letter "t." As a youngster, I used to wear a Ukrainian costume at public events (Photo 2.3), and many of my childhood friends were Ukrainian.

I loved my fifth and sixth grades in elementary school. In my homeroom, I sat by a wall of windows, under which sat a shelf of colorful books about planets and the stars. I wished I could see these through a telescope. My wish came true when I met Gary, who lived just a block away from me. He seemed to know a lot about stars already. No surprise since his father, a physics professor at the University of Alberta, was director of the university's observatory. Thus, Gary and I were able to spend Friday evenings using the observatory's telescope to look at the moon's craters, the rings of Saturn, the Great Red Spot of Jupiter, and the stars in the Milky Way. This is my first memory of being fascinated by astronomy. Sadly, about a year later, we moved to another house where I took up new hobbies, and I lost touch with Gary and the opportunity to do more star gazing.

As my school was in the University of Alberta's Faculty of Education building, we were often used as guinea pigs by the professors and teachers-in-training. The Dean, Milton LaZerte [1], liked to experiment with different ways of teaching mathematics to youngsters [2]. One day he came into our classroom and said, "We're going to do arithmetic in a new way." I was scared at first by LaZerte's abrupt, almost gruff manner, but the sparkle in his eye and his sense of humor did much to reassure us. Overall, I really enjoyed these experimental teaching classes although I remember little of the details. I think he introduced us to set theory, which became the foundation

Photo 2.3 Ukrainian children enthralled by the singer Sherril Lanyon at a concert to mark the opening of the Canada Save the Children Fund campaign in Edmonton in 1949. I'm on the left of the back row of children. From the Edmonton Bulletin, 5 February 1949

of the "new maths." I must have done well in his classes since I was singled out to be tested for my IQ. I never found out what it was.

2.2 Junior High School Years: 1951–1954

During my junior high school years, I developed close friendships with two boys. I did archery (Photo 2.4) with Bill Putnam and camped overnight by Whitemud Creek with George Bulgin, who also played in the Edmonton School Boys Band. Thanks to George, I learned to play the flute when I was 12. When George took me for a tryout in the band, Mr. Newlove, the band master, said, "I need a flute player, not another clarinetist (George's instrument). Would you be willing to learn the flute?" I was at first disappointed on hearing this, but since I really wanted to join George in the band, I agreed.

With a borrowed silver-plated flute from the band, I started private lessons with Matt Spence, a medical student who played first flute in the U of A Symphony. After 6 months of lessons, I joined the junior section of the school band in September 1953. Matt's friendly nature and high-quality flute-playing inspired me to continue with flute lessons. Since I was getting tired of my piano lessons, I was happy to quit these (although I did manage to complete a grade seven piano exam with the

Photo 2.4 Me at target practice with my homemade bow and arrows. At age 15 I won a trophy as the top Edmonton Junior Archer

Toronto Conservatory of Music in 1954). I took annual flute exams for 5 years with the Western Board of Music, which was affiliated with the U of A. I practiced a lot for these exams and always got first class grades. I wondered, might I have a career in music?

When I was 13, my sister Helen married David Osoba (also of Ukrainian descent) who was a medical student at the U of A (Photo 2.5). I was happy that she was marrying the man she had loved for the past 3 years, but sad because I was losing the only female friend that I had. I was shy with girls in junior high school and had only Helen to talk to about boy-girl relationships and feelings. I wept at the dance after the wedding ceremony.

As a young teen I continued to have many hobbies, including making tissue paper-balsa wood airplanes, building model warehouses and stations for my electric train set, collecting stamps, and doing watercolors of landscapes and buildings. To pay for these hobbies and to put money into my new savings account, I sold newspapers and greeting cards door-to-door in the neighborhood. It dawned upon me around this time that with my hobby interests and artistic skills, I could be an architect. My opening line as a neighbor responded to my knock at the door was "Would you please support my goal to be an architect by buying these Hallmark greeting cards?" However, the nearest architecture school was at the University of Manitoba, and my parents couldn't afford to send me out of town for university.

Photo 2.5 My sister Helen signing the marriage certificate in the presence of the Ukrainian Orthodox priest and husband-to-be David Osoba

In junior high school, I loved mathematics, geography and "manual training," where I made a bow and arrow holding rack and a laminated bow consisting of hickory and lemon wood. Today, my son Paul has this smoothly polished bow hanging on the wall of his basement rec room. In mathematics I found algebra challenging at first; but once mastered, I had fun with its many applications.

In June 1954, at the end of grade nine, I wrote the province-wide graduation exams. I did okay (got a mixture of A's and B's), well enough to go on to high school, but the results did not suggest that I had an aptitude for any special subject. During summer, I ran into my home room teacher, Mr. Lynn. "Did you get all A's?" he asked. He was surprised and disappointed when I said no. I was not always good at writing exams, especially those provincial ones which covered material we never saw in class.

2.3 High School Years: 1954–1957

I did my first year of high school (grade 10) at the University High School, in the same building where I also did my last 2 years of elementary school. However, the south side of Edmonton was rapidly expanding, and a new high school was built to accommodate the influx of teenagers to this region. Thus, for grades 11 and 12, I moved to Strathcona Composite High School (SCHS) for my last 2 years of high school (1955–1957).

SCHS had both academic and technical programs, such as automotives, woodworking, electronics and plumbing. Although I was in the academic stream, I was delighted to be allowed to take woodworking as an elective. I made a ping pong table which later followed me to Vancouver and Montreal. I was also encouraged to participate in the sports programs. Not being good at team sports, I chose to take up badminton and swimming.

Outside of school, I continued with my flute lessons, now under the tutelage of Joan Pecover, a talented flutist with the Edmonton Symphony. At age 16, I upgraded to a professional-level Haynes flute made in Boston, thanks to the generosity of my mother, who paid $ 350 of the $ 400 price tag. That was a lot of money in those days, but very much worth it. Now, seven decades later, I can still get a warm sound from this solid silver gem.

While in high school, I continued to be active in the Canadian Boy Scouts movement. I had a most dedicated Scoutmaster, Dr. Donald Robinson, who was a chemical engineering professor at the U of A. He took us camping in both summer and winter (freezing in a tent at −20 C in the northern Canadian woods was no fun!), and he taught us leadership by example. He also recommended that I give an appreciation speech at the Queen's Scout Recognition Ceremony in October 1954, when I received my Queen's Scout Certificate. The next year a select few of us (Photo 2.6) were lucky enough to be chosen to attend the International Scout Jamboree in Niagara-on-the-Lake, Ontario.

Near the end of my high school years, I became less shy with girls. This was partly due to my participation in the Canadian Ukrainian Young Peoples club, going by the acronym CYMK in Ukrainian. In addition to barbeques, softball games, and dancing and skating parties, we participated in choir and handicraft activities (Photo 2.7). When I got my driver's license at 16, I started dating. I took Kathy Rusnack to the movies on my first date. She was vivacious as ever when I met her again in Calgary in 2003, where I gave a talk about my recent lecture tour in Ukraine.

In high school, I had excellent teachers in mathematics (Miss Scott), physics and chemistry (Mr. Radomsky), French (Miss Woods) and English (Miss Brown). I got top marks in chemistry and very good marks in the other subjects, high enough to be one of five in the 225-student graduating class who received an academic award. Remarkably, three of us five went on to study mathematics at the U of A. In the province-wide graduation exams, however, my top mark was in French; English and mathematics were near the bottom! Not surprising that I thought engineering was out for me as a career. Today, seeing my son as a successful civil engineer in British Columbia, I think I could have enjoyed being an engineer as well.

For extracurricular activities, I teamed up with music teacher E. C. Mayes to form a school orchestra. We were a strange mixture, with far more wind and brass players than strings players (Photo 2.8). Mozart would have shuddered hearing our rendition of the Magic Flute overture. But we had a lot of fun, giving about 10 performances over the next 2 years. At some of the concerts, I was asked to conduct the orchestra, which I loved.

The last fun activity I did before graduation was organizing the Graduation May Ball with a team of five friendly girls (Photo 2.9). Clearly by now I felt very comfortable being with members of the opposite sex. At the Ball, we danced to Frank McCreavy's Big Band Orchestra, and my Dance Program (Fig. 2.1) was full, with

Photo 2.6 Scouts from the 41st Edmonton troop who went to the International Scout Jamboree in summer 1955. I'm in the middle of the second row

Lorraine Stepchuk being my date and favorite dance partner. We renewed our friendship in Edmonton a few years ago at a high school reunion. "Do you remember that a group of us went to the Purple Lantern for Chinese food after the ball?" she asked. We teenagers were always hungry.

I was an excellent student in high school (I was a co-recipient of the Scarborough Memorial Trophy for outstanding service and academic achievement (Photo 2.10), but I still had a hard time deciding what I would take at university, much to the frustration of my parents. I was envious of my cousin Allan Trojan in Vancouver who knew at age 16 that he wanted to be a mathematician (Photo 2.11). At age 23, Allan completed a PhD in mathematics at MIT and became one of the youngest professors at McGill University in Montreal.

After a thoughtful consideration of the list of professions that my parents suggested (law, medicine, teaching, engineering, business, and dentistry), I finally decided that

Photo 2.7 A group of CYMK members from St. John's Ukrainian Orthodox Cathedral preparing for a handicraft exhibit and tea in 1956. Left to right: Lydia Bayrack, Pat Proniuk, Bill Charnetski (standing), Evan Verchomin, and myself, demonstrating the burning of Ukrainian designs on a wooden plate. Bill, whom I met when I was six, is my longest standing friend; we met last in Niagara-on-the-Lake, Ontario in 2023. Material (4 June 1955) republished with the express permission of Edmonton Journal, a division of Postmedia Network Inc

Photo 2.8 The Strathcona composite high school orchestra (1956–1957 tricolor yearbook), organized by me and E.C. Mayes in 1955. Lynne Newcombe (second from left in middle row) was a very talented pianist with whom I later gave many flute-piano recitals at the University of Alberta. I'm fifth from the left in the second row, wearing a bow tie. Mr. Mayes, our conductor, is in the top left corner. Permission granted from the Strathcona composite high school

Photo 2.9 Me (president of the Strathcona composite grad executive) with the friendly five: L to R, Audrey Pystoski (Class Historian), Carolyn Smith (vice president), Valerie Logan (treasurer), Doreen Draper (Valedictorian) and Barbara Logan (secretary). On the far right, staff advisor Maurice Rookwood looks on. Material (16 February 1957) republished with the express permission of Edmonton Journal, a division of Postmedia Network Inc

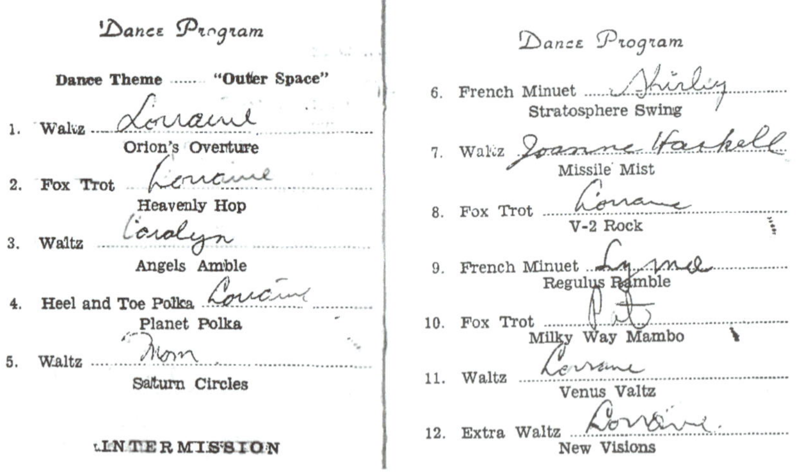

Dance Program

Dance Theme ...... "Outer Space"

1. Waltz Lorraine
Orion's Overture

2. Fox Trot Lorraine
Heavenly Hop

3. Waltz Carolyn
Angels Amble

4. Heel and Toe Polka Lorraine
Planet Polka

5. Waltz Mom
Saturn Circles

INTERMISSION

Dance Program

6. French Minuet Shirley
Stratosphere Swing

7. Waltz Joanne Haskell
Missile Mist

8. Fox Trot Lorraine
V-2 Rock

9. French Minuet Lynne
Regulus Ramble

10. Fox Trot Pat
Milky Way Mambo

11. Waltz Lorraine
Venus Valtz

12. Extra Waltz Lorraine.
New Visions

Fig. 2.1 Dance program for the graduation May Ball. Lorraine, an excellent dancer, looked stunning in a pink gown

Photo 2.10 Strathcona composite high school (SCHS) award ceremony where the major trophies were awarded (10 June 1957). I'm second from the right with my hand on the Scarborough memorial trophy, which I shared with Murdith McLean. On the far left, Bob MacLeod is holding the Wees cup, for top marks in the graduating class. Bob and I both ended up in honors mathematics at the University of Alberta. Material republished with the express permission of Edmonton Journal, a division of Postmedia Network Inc

I would enroll in the pre-dental program at the University of Alberta (Photo 2.12). I thought this would provide for a good living and would make excellent use of my handicraft skills.

2.4 University of Alberta: 1957–1961

I did not last long in the pre-dental program. Dissecting twitching frogs in the zoology lab put me off from further work using scalpels and other sharp objects. Fortunately, I had an inspiring calculus professor, Reginald Jacka, who brought to life the wonderful world of mathematics and its applications. I even got 100% on a midterm exam! He recommended to the math department head, E.S. Keeping, that I should switch to an honors math program in my second year. "Because of your practical inclinations, I think you would enjoy our honors applied mathematics option," Keeping suggested. Hence for the next 3 years, I took 15 courses in physics and pure and applied mathematics.

I had several excellent professors, many of whom became FRSC (Fellow of the Royal Society of Canada). Lee Lorch taught me differential equations and complex analysis, subjects which I enjoyed even though his classes were often poorly prepared.

Photo 2.11 Cousin Allan Trojan, who solved math problems which stumped his grade 12 teachers at North Vancouver High School (November 1956). Material republished with the express permission of Vancouver Sun, a division of Postmedia Network Inc

BETH MURRAY
Beth's destination is U. of A. and then on to Social Welfare work. Beth has a great affection for swimming, skating, piano and long weekends. Being Secretary for the Red Cross and Y-Teens helps keeps her busy this year.

LAWRENCE MYSAK
The President of our Grad Class takes a keen interest in music. He has won two academic awards and a U.N. scholarship to Banff in 1956. Likes dancing, badminton and the flute and plans to attend Varsity for Dentistry.

Photo 2.12 My graduation picture in the SCHS tricolor yearbook for 1956–1957. Beth Murray became a well-known social worker in Edmonton

Leo Moser gave stimulating lectures on finite mathematics and abstract algebra; he never seemed to need any notes for his lectures and told lots of mathematical jokes. Don Betts taught me mathematical physics in which I got top marks, even though I never fully appreciated the significance of a Green's function until I had to teach this subject at UBC. My favorite professor was the then young Werner Israel, who taught a graduate course on tensor analysis and general relativity in my last year. His excellent preparation of the classes and smooth delivery of the lectures convinced me that relativity would be a good area to do research in, which I did in Adelaide,

Photo 2.13 Me as flautist at age 19 with the University of Alberta Symphony, holding my silver Haynes flute

Australia (Chap. 5). Israel went on to a very distinguished career in his field. He became a Fellow of the Royal Society (FRS) and a collaborator of Stephen Hawking at the University of Cambridge.

Alongside my mathematics and physics studies, I continued with lessons in flute and music theory. In 1958 I completed the practical component for a performance diploma in flute, and by 1960, I finished all the theoretical components for this diploma (music history, counterpoint and harmony). After spending a summer at the Aspen Music School in Colorado (Chap. 26), I thought that music might be an alternative career for me. I was highly appreciated as both a player (Photo 2.13) and assistant conductor with the University of Alberta Symphony (Fig. 2.2). Also, before graduation I had served as president of both the University Symphony and the University Music Club, where I gave many recitals with Lynne Newcombe as pianist. When I asked music professor Richard Eaton for advice on a musical career, he replied, "Lawrence, do you want to be poor? If not, then stick to your mathematics." And hence, like my cousin Allan, I became a mathematician.

In the fall of 1960, I was awarded a Rotary Fellowship for International Understanding for study at the University of Adelaide (Photo 2.14), where I planned to do research in general relativity in collaboration with George Szekeres (Chap. 5). I graduated in May 1961 (Photo 2.15) and then spent the summer in Victoria working at the Pacific Naval Laboratory under the supervision of John Weaver (Chap. 3). Both Szekeres and Weaver were superb mentors and became lifelong friends. In fall 1961 I had my first experience of university teaching (Chap. 4). After a bumpy start, I really felt at home as a university lecturer and convinced myself that an academic appointment would be my long-term goal.

As I boarded the plane for Australia in January 1962, I was excited but feeling lonely. I was going to miss the marriage of my best friend Bill (Photo 2.16) and was

Fig. 2.2 Edmonton journal write up (26 February 1959) of the February 1959 concert of the University of Alberta Symphony. Material republished with the express permission of Edmonton journal, a division of Postmedia Network Inc

University Symphony Gives Lively Concert

A student conductor, second year honors applied mathematics student Lawrence Mysak, and a piano soloist, music student June Daley, supplied the highlights of an evening of diversified music at the University of Alberta Symphony Orchestra's annual concert in Convocation Hall Thursday.

Usually a flautist with the orchestra, Mr. Mysak stepped to the podium to conduct the orchestra through melodies from Richard Rodgers' "The King and I."

These modern selections not only added variety to the concert but gave an opportunity to hear, in public performance, music of a more current semi-popular nature played by a symphony-sized orchestra.

THOROUGHLY AT HOME

This young conductor and the orchestra were thoroughly at home with the selections from "The King and I." Their interpretation effectively showed the fullness of expression that can be given to familiar melodies.

L. A. MYSAK

Fellowship Awarded To City Student

Photo 2.14 Public announcement (15 December 1960) of my being awarded a Rotary Foundation Fellowship for International Understanding for graduate studies in Australia. Material republished with the express permission of the Edmonton Journal, a division of Postmedia Network Inc

Photo 2.15 In my graduation gown with mom and dad. May 1961

leaving many other friends behind. During my university years I dated several girls but never found the right one. Maybe in Australia I would?

Photo 2.16 Bill and Emily Charnetski on their wedding day in September 1962

Acknowledgments I thank Janet Boeckh and Joseph Schmidt for their insightful comments, and I very much appreciate having the newspaper clippings and old photos that my mother saved. I am grateful to Olivia Marino for inserting the photos.

References

1. Chalmers JW (1978) Gladly would he teach: a biography of Milton Erza LaZerte. Alberta Teachers' Association Trust, Edmonton, AB
2. LaZerte M (1927) A study of the methods used by elementary school pupils in solving problems in arithmetic. PhD thesis, University of Chicago

Part II
Life as a Mathematician, Climate Scientist and Professor in Canada, Australia, the United States, and Europe

Chapter 3
Dr. John Weaver, Summer 1961 Supervisor

Abstract In the summer of 1961, I worked as a student research assistant at Victoria's Pacific Naval Laboratory (PNL), a unit of the then Canadian Defence Research Board. Although I had never studied fluid mechanics as an undergraduate student at the University of Alberta, I was assigned the problem of describing mathematically the flow of a fluid past a cone that was used to measure turbulence in the ocean. It was a challenging problem as I first had to learn the basics of fluid mechanics and the concept of a boundary layer. But I had excellent guidance from my supervisor, John Weaver who had recently joined the research staff of PNL. This project was a turning point for me as when I later went on to do a PhD in applied mathematics at Harvard University, I decided that I would like to do a PhD thesis in some area of fluid mechanics.

"John, this is Lawrence from Montreal. How are you getting along on the balmy west coast?" I ask over the phone.

"Just fine, thanks," John replies. "I'm still enjoying morning walks with my fellow seniors. Keeps me in shape."

"That's great. Janet and I are visiting Victoria and wondered—are you free for coffee tomorrow afternoon?"

"Certainly," he says. "How about at 3 p.m.? I'll pick you up where you're staying."

In December 2017, Janet and I were in Victoria, British Columbia for Christmas with her son Rob and family. Whenever I visit Victoria, I always like to see John and his charming Ukrainian wife, Ludmilla. John is the consummate Englishman—polite, cultured, and an accomplished scientist; he's also keen on soccer. More than half a century ago, he was a game-changer for me when in summer 1961, I worked in Victoria at the Pacific Naval Lab (PNL), a federal research institution.

I had just graduated from the University of Alberta, and the PNL job gave me insight into a possible future career as a scientist. John Weaver, on the other hand, had just finished a PhD in geophysics from the University of Saskatchewan and was starting his scientific career at PNL in 1961.

When I arrived at PNL, I started working for oceanographer Dr. Harold Grant. I computed energy spectra of ocean temperature and velocity measurements using a

L. Mysak, *Adventures in Climate Science, Ocean Waves, and the Flute*, Springer Biographies, https://doi.org/10.1007/978-3-032-19848-8_3

slide rule. This sounds complex, but in fact I just multiplied numbers in one column with the numbers in another column and then added. I was bored, but I had to make a presentation about my project to the staff. I said diplomatically, "Surely, there must be another project in which I can use my advanced maths and physics."

Dr. Grant looked upset. But to his credit, he asked his junior associate Blyth Hughes and the young Dr. Weaver, "Do you think you could find a fluid mechanics problem that Mysak might tackle for the rest of the summer?" I was grateful to Dr. Grant for understanding my dilemma and allowing me to take on a project of more mathematical content.

Next day, they suggested I try to solve the equations that describe the flow past a probe used in ocean turbulence measurements. Since I had never taken a course in fluid mechanics, I pored over the treatise *Boundary Layer Theory* [1] for help to solve this problem. While Hughes was my immediate supervisor, it was Dr. Weaver who tutored me in fluid mechanics, one of his passions. Step by step, he led me through the mathematics that resulted in an approximate solution to the assigned problem. I was thrilled with the project and wrote two reports at the end of the summer [2, 3]. Dr. Weaver also gave me many tips on how to do research. "You have to always look for another approach if one fails," he said. At the end of the summer, I thought that a problem in fluid mechanics would be a fascinating topic for a PhD thesis.

Dr. Weaver also added an extra dimension to the lives of us summer students working at PNL. During evenings Weaver taught us the finer skills of soccer (football in Europe and elsewhere outside of Canada and the US), and on weekends we played friendly games with other regional teams. That fall Ludmilla gave birth to a son, Andrew.

After my summer in Victoria and a year of graduate study in Australia, I entered Harvard in fall 1963 for my PhD in applied mathematics. Dr. Weaver was delighted that I asked him for a letter of reference for Harvard—it must have been a good one, because in addition to acceptance for the PhD program I received a scholarship which covered my tuition. At Harvard, I returned to the topic of my PNL research, fluid mechanics. I also enjoyed studying aerodynamics and, new to me, geophysical fluid dynamics (GFD). My PhD research focused on the GFD topic of continental shelf waves [4].

Beginning in 1984, I tried to inspire another young student, in the manner that John had inspired me two decades ago. During 1984–1987, I supervised the research of John's brilliant son Andrew Weaver who was doing a PhD in applied mathematics at UBC. It was a pleasure for me to watch Andrew, with his intriguing British/Ukrainian temperament, go on to an outstanding career as a professor of oceanography and climatology at the University of Victoria. For 5 years (2015–2020), he was the very busy leader of the BC Green Party and an elected Member of the Legislative Assembly in British Columbia. Consequently, when I visited Victoria to visit family during this time, I didn't see my former student, but instead saw his father, my friend John, who was my inspiration in 1961.

John moved to the University of Victoria in 1963 and retired in 1998, but our friendship continued to develop. We had fun serving on research grant selection committees together, and colleagues told me that he was a highly respected Dean of

Science before retirement. As we are now both retired, we love to chat about former students and colleagues and discuss the challenges and frustrations of university politics. I hope we continue to meet for coffee for a long time to come.

3.1 Epilogue

In the fall of 2024, John started to have serious health problems, and in May 2025, I learned that he passed away peacefully on 26 April 2025, with the immediate family at his side. He was 92. John was such an important mentor to me early in my career, and I know I will always have the memories of our many past interactions.

This chapter is an update of an assignment which I was given in the Thomas More Institute (TMI) Memoir Writing Course that I took in winter 2018 in Montreal. I am very grateful to Pauline Beauchamp and Karen Nesbitt, the course instructors, for their encouragement, feedback, and suggestions for the improvement of this short memoir.

References

1. Schlichting H (1960) Boundary-layer theory. McGraw-Hill, New York
2. Mysak LA (1961) On the velocity profile for flow past a hemispherically-tipped cone. Pacific Naval Laboratory (PNL) Note 61–15 Defence Research Board, Esquimalt BC, pp 34
3. Mysak LA (1961) Determination of the boundary layer velocity profile for flow past a hemispherically-tipped cone using an approximate method. PNL Note 61–16, Defence Research Board, Esquimalt BC, pp 9
4. Mysak LA (1967) On the theory of continental shelf waves. J Mar Res 25:205–227

Chapter 4
From Disaster to Enlightenment: How I Learned to Teach

Abstract In September 1961, I was appointed as a sessional lecturer in the Department of Mathematics at the University of Alberta. Early one Saturday morning, I was asked to fill in for a "no-show" new faculty member who had been assigned to teach second-year calculus and direct a statistics laboratory the next Monday. I was only 21 at the time and had never lectured in a large class. After a stressful first lecture, I quickly learned how to relax and interact with the students. Soon I felt at ease teaching at the university level and realized that this is what I wanted to do as a career, be a university professor.

In September 1961, I became a visiting graduate student in the Mathematics Department of the University of Alberta (U of A), where I had just graduated with an honors degree in applied mathematics. I took this visiting position to keep my brain in gear while waiting to fly off to Australia in January to do my masters (Chap. 5). This brain work involved auditing a few graduate mathematics courses given by some of my favorite professors. The department also gave me a small office where I could study the notes I took while auditing these courses.

While sitting in my office one Saturday morning (yes, there were classes on Saturday then), Reginald Jacka, one of my former inspiring professors, came rushing in. "The department head, Max Wyman, wants to speak to you," he told me. I was puzzled by this sudden request. Jacka solemnly led me down the hall to the departmental office to see Wyman.

"Lawrence, we have a huge problem coming up next week," Wyman said. "We have a "no-show" new faculty member who was supposed to start lecturing to a second-year calculus class and direct a statistics lab on Monday. Can you step in and take over these tasks for us?" he asked.

I was stunned. I was only 21 at the time and had never lectured in a large class of students. However, I had given presentations at our undergraduate math and physics club, and the word had gotten out that I could explain solutions to mathematical problems very well. That was good enough for Wyman. "You have the weekend to prepare before you meet your first class on Monday," he said. "And by the way, we'll

L. Mysak, *Adventures in Climate Science, Ocean Waves, and the Flute*, Springer Biographies, https://doi.org/10.1007/978-3-032-19848-8_4

pay you $1000 (over $10,000 in today's dollars) for your efforts until you leave for Australia in January."

For the rest of the day, I feverishly prepared notes on what I planned to do on the first day of class, namely, a review of the first-year calculus material. I was less stressed about directing the statistics lab, as two summers ago I worked for the Canadian Defence Research Board doing statistical analysis of geophysical data [1].

Dressed in a dark suit and tie and with nerves on edge, I walked into a classroom of over 100 students eagerly waiting to scrutinize the new calculus professor that Monday. I could see immediately that many in the room were older than me! After I gave them my name and office number, I quickly turned toward the blackboard and started racing through my review of freshman calculus. I repeatedly filled the blackboards with equations and symbols, and I never looked back at the students until the end of the hour. I don't know what was shaking more, my legs or my hand holding the chalk. Miraculously, I got through the hour still standing and dashed out of the room before any student could start asking questions. I had no idea as to how the students reacted to what I threw at them in my first "lecture."

Before giving the next class, I got to the classroom early and asked an entering student whether they found my first lecture okay. "You terrified us," she warned. "At the rate you went on Monday, you'll finish the course by November!" Their message: please slow down today.

And this I did. I faced the class at the outset and mentioned the topic that I would cover today and invited them to ask questions at any time, even though I feared losing control. What if I couldn't answer their questions? But I sensed that this invitation made them more at ease, and I began to understand what the roadblocks to their learning might be. As I developed the day's topic on the blackboard and answered their questions, I started to relax. I also got to know the names of a few students. Soon I was enjoying the process of university teaching, and I realized that this interactive approach worked very well. By the end of the hour, I could see smiles of understanding on the students' faces.

After the second class I hung around to talk to a few students about their programs and career plans. I learned that several of them aspired to be teachers of science or mathematics. They were especially interested in any applications of the type of mathematics that they had learned today. As a graduate in applied mathematics, I would have no difficulty in providing such applications.

As the semester progressed, I enjoyed the calculus classes increasingly, as did the students. As I gained experience, I learned how rewarding teaching could be, and this brought me closer to my parents who were accomplished teachers. I soon felt that a career in university teaching was in the cards for me.

Surprisingly, no one in the department ever seemed to check up on my teaching performance that fall. But it must have been more than okay, because I was warmly applauded by my students when I finished my teaching contract in January, and a couple of years later, I was hired back by the U of A to teach the "new mathematics" to mature math teachers in the summers of 1963, 1964, and 1967. By the time I started my first job as an assistant professor of mathematics at UBC in September 1967 (Chap. 10), I had made the transition from a formal lecturer to a skilled teacher.

In summer 1986, I moved to McGill University in Montreal to begin my career in climate science (Chap. 15). But I didn't leave my enthusiasm for mathematics and its applications behind. At McGill I was based in the Department of Meteorology (now Atmospheric and Oceanic Sciences), and I stopped teaching mathematics, although I was happy to be made an associate member of the Department of Mathematics and Statistics. Sophisticated mathematical techniques and statistical methods are needed to analyze the vast amounts of environmental data that are being collected by various national and international agencies. I first learned that these data show evidence of different types of climate change, both in the present and in the past. Secondly, I started to develop physics-based mathematical models of the earth system to explain these climate changes and to forecast future climate scenarios. To teach the students about these exciting scientific developments, I created a new course in climate and paleoclimate dynamics based on the data in Ruddiman [2] and supplemented by the application of different classes of climate models. These included Boolean delay equations [3], box (dynamical system) models [4], two-dimensional models [5], and Earth system Models of Intermediate Complexity (EMICs) [6]. This course attracted outstanding graduate students to my research group, many of whom came with strong mathematical backgrounds (Chap. 24). Thus, not only did I begin teaching mathematics and statistics six decades ago at the U of A, but for the remainder of my academic career, I continued using mathematics and statistics to try and explain the nature and causes of the significant changes in our past and present-day climate.

Acknowledgements I thank Janet Boeckh and Josef Schmidt for their helpful comments.

References

1. Mysak LA, Jacobson TV (1960) A survey of geomagnetic impulses observed during solar cycle 1949–1959. Memorandum, Defence Research Telecommunications Establishment, Defence Research Board, Ottawa, p 49
2. Ruddiman WF (2008) Earth's climate: past and future. WH Freeman, New York
3. Darby MS, Mysak LA (1993) A Boolean delay equation model of an interdecadal Arctic climate cycle. Clim Dyn 8:241–246
4. Carozza DA, Mysak LA, Schmidt GA (2011) Methane and environmental change during the Paleocene-Eocene thermal maximum (PETM): modeling the PETM onset as a two-stage event. Geophys Res Lett 38:L05702. https://doi.org/10.1029/2010GL046038
5. Stocker TF, Wright DG, Mysak LA (1992) A zonally averaged, coupled ocean-atmosphere model for paleoclimate studies. J Climate 5:773–797
6. Claussen M, Mysak LA, Weaver AJ, Crucifix M, Fichefet T, Loutre M-F, Weber SL, Alcamo J, Alexeev VA, Berger A, Calov R, Ganopolski A, Goose H, Lohmann G, Lunkeit F, Mokhov II, Petoukhov V, Stone P, Wang Z (2002) Earth system models of intermediate complexity: closing the gap in the spectrum of climate system models. Clim Dyn 18:579–586

Chapter 5
George, Paul, and My Erdős Number of 2

Abstract In January 1962, I arrived in Adelaide, South Australia to begin research for an MSc in mathematics at the University of Adelaide. My research supervisor would be the distinguished Hungarian Australian mathematician George Szekeres, and my topic of research was going to be related to Einstein's theory of general relativity. While doing research on a problem involving a 'black hole', I learned that Mr. Szekeres was also an early collaborator of the itinerant and prolific Hungarian mathematician Paul Erdős, who had over 450 co-authors. Each one of them has been assigned an Erdős number of 1. Since George and I later published a paper together in the Canadian Journal of Physics, I have been assigned an Erdős number of 2. My colleagues in climate science are astonished by this strange recognition!

In late January 1962, it was −30 C in Edmonton. I was so lucky to escape this frigid winter by traveling to Australia, where I was soon to begin graduate studies in mathematics at the University of Adelaide. I was fulfilling a dream that started 18 months ago in Ottawa, where I was working for the summer in geophysics for the Defence Research Board of Canada. My office mate Thor Jacobson and I often talked about possible universities for graduate work.

"I'd love to go to Australia," Thor said. "The weather is great, and the women are beautiful."

"Great idea," I replied, and "I understand there are very good universities in Australia."

Later that summer, I applied for various scholarships for study abroad. In the fall of 1960, I was thrilled to be awarded a Rotary Fellowship for International Understanding for a year of study in Australia. I chose to attend the University of Adelaide for two reasons—I could complete an MSc there in one year, and I found a professor at the university, George Szekeres, who was willing to take me on as a research student to work on mathematical problems in Einstein's theory of general relativity, my scientific passion at the time.

During the long flight to Australia, I was both excited and nervous. I had just turned 22, and I had never traveled outside of North America. I didn't know a soul in Australia. My mother really thought I would never return to Canada. "You will meet

L. Mysak, *Adventures in Climate Science, Ocean Waves, and the Flute*, Springer Biographies, https://doi.org/10.1007/978-3-032-19848-8_5

this attractive Australian girl and stay down under," she warned. As it turned out, I did meet several lovely young women in Australia. But in the end, I never found the 'right one' and thus her worries were for nothing.

Upon exiting the plane in Adelaide, I was hit by the humidity and heat—it was nearly 35 C. I was also overwhelmed to be greeted by members of the Adelaide Rotary Club, a mathematical physics graduate student from the University of Adelaide, and the press. Because I was supported by this prestigious Rotary Fellowship, this was big news locally—it was the first time a Rotary Fellow from abroad had come to study at the University of Adelaide.

After settling into St. Mark's College, my residence for the year, I made my way to the Department of Mathematics at the 'uni'. I was anxious to meet my supervisor. However, I discovered that most professors were away for the summer, including Mr. Szekeres. We did not meet for another month. By this time the weather had finally cooled down.

George was addressed as *Mr.* Szekeres because he did not have a PhD. His only academic qualification was a Diploma in Chemical Engineering, which he had obtained in the early 1930s from the Technical University of Budapest. While in Australia, I gradually learned more about his unusual background.

During the 1930s George's father owned a leather factory in Budapest, and it was expected that as a trained engineer and an only son, he would one day take over the family business. However, George was enthralled with mathematics, and while studying and later working in engineering, he often hung out with a group of young mathematicians in Budapest. They liked to solve challenging problems and propose conjectures in a wide range of mathematical subjects, including number theory, algebra, combinatorics, geometry, graph theory, and mathematical analysis. This gang of about a dozen mathematicians, who were all Jewish, met regularly at the foot of a statue in Budapest's City Park. The leaders of the group were George, Esther Klein, Pál Turán, Tibor Gallai, and Paul Erdős, and over the years they published many papers together. George, Pál and Paul can be seen playing ping-pong in Photo 5.1. With the rise of anti-Semitism in central Europe at the time, the group wisely dispersed just before World War II. For example, George and Esther (who married in 1937) fled to Shanghai, and Erdős escaped to Princeton in the USA.

Even while working as a chemist in a leather factory in Shanghai all through the war years, Szekeres published several papers in various fields of mathematics. This remarkable accomplishment enabled him to obtain a lectureship in mathematics at the University of Adelaide in 1948. Once settled in Australia, he thrived on having the opportunity to focus on teaching and doing further research in mathematics. By 1962, when I arrived in Adelaide, he was regarded as one of Australia's leading mathematicians. George was then 50. Later, I wanted to hear more about the fate of the others in the mathematics group whom he left behind in Budapest in the late 1930s. Had any one of them become a famous mathematician? Yes, Paul Erdős!

In 1948, after 10 years traveling about in the USA, Paul Erdős returned to Hungary at age 35 to continue his research in mathematics. By then he had a wide network of colleagues around the world with whom he visited often and published papers. Paul lived and breathed mathematics. He idolized his mother and never married.

Photo 5.1 Erdős (left), Szekeres (middle), and Turán playing ping-pong in 1958. Szekeres and Turán were the first and second mathematicians to have been assigned an Erdős Number of 1. Picture from [2], courtesy of Janos Pach

Further, his style of working and publishing was unique. Not having a permanent position Erdős lived an itinerant life, traveling from city to city and begging his mathematician friends for short-term accommodation. Upon arrival at airport X, where he was invited to give a seminar at university Z, he would phone up colleague Y to say, "Consider the following sequence of prime numbers $p_1, p_2, \ldots$.. I conjecture that ...." Then Paul would add, "By the way, I need a place to stay for a few nights. During my stay I think we can prove this conjecture together." Later they did indeed succeed in outlining a proof. However, Paul would leave colleague Y to write up the joint paper and then be off to his next destination ZZ and speaking engagement.

By the late 1960s a concept was born to describe the extent of Erdős's collaborations: the Erdős Number [1]. Anyone who has published a paper with Paul is said to have an Erdős Number of 1. A person who is a collaborator of that person is said to have an Erdős Number of 2, and so on. At a mathematics party, a good pick-up line is "What is your Erdős Number?".

During his lifetime of 83 years (1913–1996), Erdős published over 1500 papers and collaborated with over 450 different mathematicians. Remarkably, my supervisor George Szekeres was the first mathematician to publish a paper with Erdős and thus was the first to have an Erdős Number of 1. This paper [3] has been cited over 450 times. When I met George in 1962, I knew nothing about Erdős, his fame and eccentricity, or the extent of his network. Nevertheless, George often spoke of Erdős with great respect and fondness, and Erdős often visited the Szekeres family in Australia. "Lawrence," George said to me in Adelaide, "you will have to meet Erdős one day, to see for yourself the nature of true genius." This meeting did not happen for another 12 years.

My year in Australia was memorable. I got on very well with George and his charming wife Esther and their young daughter Judy; I was treated as part of their family. The Szekeres's son Peter had just gone off to London to do a PhD in mathematical physics, and I was a sort of surrogate son. George was always available in his office to discuss my research, which I found very challenging. In addition, George had a very strong accent, and it took me several months before I really understood his Hungarian English *and* his advice. George and I played chamber music together (violin and flute respectively), went hiking with other math students in the Adelaide Hills, and had many lively debates in the mathematics coffee room. After about eight months, I had enough results from my research for an MSc thesis, and in early 1963 we submitted a paper for publication to a high-level physics journal which published papers on general relativity. After many months, we finally received a reply from the editor along with a referee's report. The main conclusion: "While this paper has some merit, it is not of sufficient merit to warrant publication in the Journal of Mathematical Physics." Our paper was rejected! And I felt terrible. However, the referee did make some useful suggestions to improve the paper.

I received this bad news in the fall of 1963. Since I had just started a PhD in applied mathematics at Harvard University and was up to my ears in coursework, I had little time to revise and re-submit the paper to another journal. However, I was able to do this in August 1965, by which time all my courses and PhD qualifying exam were completed. Happily, the paper was accepted and published the next year in the Canadian Journal of Physics [4].

Much later, I learned that this journal publication [4], my first, earned me an Erdős Number of 2. There are over 5000 mathematicians and scientists with an Erdős Number of 2 versus 450 with an Erdős Number of 1. Perhaps it is realistic to say an Erdős Number of 2 is not such a big deal! But according to a website of Erdős, over 200,000 mathematicians have an assigned Erdős number.

My paper with Szekeres dealt with the mathematical nature of the so-called Schwarzschild singularity which appears in the solution of Einstein's gravitational field equations for a sphere. Since the singularity at the Schwarzschild radius can be transformed away using the Kruskal-Szekeres coordinates (used in [4]), we asked the question whether distant superimposed gravitational fields could alter this feature. If these fields were static, the answer is no; if time-dependent, however, we showed that the singularity is changed into a coordinate-independent singularity. Today we know that this singularity is very relevant to our understanding of black holes, which are locations in outer space where matter is highly compressed into a region inside the radius associated with the Schwarzschild singularity. Light cannot escape from a black hole, a term coined by Princeton physicist John Wheeler.

In 1971 I met Stephen Hawking at the University of Cambridge where I was spending a sabbatical year (Chap. 11). By then Hawking was a leading expert on black holes. One day he said to me, "Oh, so you're one of the authors of that 1966

Photo 5.2 Paul Erdős at age 80. Photo by George Csicsery for his film *N is a Number: A Portrait of Paul Erdős* (1993).

paper in the Canadian Journal of Physics. It's an interesting paper, but you could have made much more out of the physical implications of your results." Obviously, he was not very impressed with the paper. But I didn't mind because by then I had long left the field of general relativity, having found the topic too abstract for my liking. I was now doing research in physical oceanography. Nevertheless, I like to think that my first refereed paper did encourage others to seriously probe into the astrophysical significance of the Schwarzschild singularity.

At the 1974 International Mathematics Congress in Vancouver, BC, I met Erdős (Photo 5.2) over lunch with Szekeres and my then wife Mary at the UBC Faculty Club. It was a strange experience. Erdős was continuously waving his hands about as he talked, jumping from topic to topic. "He was like a large skinny bird flapping about with too much energy," Mary said. Three years later, Erdős and I happened to meet at University College London where I was invited to give an applied mathematics seminar on continental shelf waves (Chap. 7). He remembered me immediately as an oceanographer who was a former research student of Szekeres. He then asked, "What would happen if you dropped an atomic bomb offshore of a large coastal city?" He was surprised that I had not thought about this problem and its fate for humanity.

The Szekeres family and I kept in touch for 40 years, from the early 1960s until the early 2000s. We exchanged annual Christmas letters, and we visited in each other's homes, both in Australia and in Canada. George moved to Sydney in 1964 to take up the Foundation Professor of Pure Mathematics at the University of New South Wales, and my then-wife Mary and I visited his home in the distant Sydney suburb of Turramura several times after this. Their home was surrounded by fragrant eucalyptus trees and overlooked a lush ravine. George's study, in the upper floor of an architect-designed addition to the back of the house, had a spectacular view of this greenery.

I once considered accepting a senior research position at a CSIRO oceanography lab in Tasmania, in the mid-1980s (Chap. 15), which would have delighted both George and Esther. But instead, I moved to Montreal to take up a newly created climate chair at McGill University. Although George retired in 1976 at age 65, he and his wife continued to travel the world attending conferences and visiting mathematical colleagues. I arranged for him to give a mathematics seminar at McGill in 1987, and I hardly understood a word of it. But his talk certainly intrigued the audience.

In 2004, at age 93, George lost his driver's license, and he and Esther were forced to leave their beautiful home in Turramura. They returned to Adelaide, to live in a nursing home and be near family. They passed away within an hour of each other on August 28, 2005. George was 94 and Esther 95. They had an amazingly long and successful marriage.

Over his lifetime, George was honored and widely appreciated in Australia and elsewhere for his contributions to mathematics and education. He received an honorary doctorate from the University of New South Wales when he retired. A Festschrift was organized for his 90th birthday by the Australian Mathematics Society (which he helped found), and shortly after that, the "George Szekeres Medal for Mathematics" was created by the Society. George was also made a Member of the Order of Australia. I often think about George and Esther even now, remembering them for their wonderful support and advice when I was a young student in faraway Adelaide, and for their friendship and love over the decades since the 1960s. After George retired, I encouraged him to write his autobiography. "No way," he replied, "I still have too much mathematics to do."

Acknowledgements I thank my wife Janet Boeckh for her careful review of this chapter and her constructive suggestions for its improvement, and Nevein Gamal for help with the photography and proofreading. Much of the above detailed information about Paul Erdős and his relationship with George and Esther Szekeres comes from the 1998 book by Bruce Schechter [2]. This biography of Erdős is a delightful read.

References

1. Goffman C (1969) And what is your Erdős Number? Am Math Month 76:791

2. Schechter B (1998) My brain is open: the mathematical journeys of Paul Erdős. Simon and Schuster, New York
3. Erdős P, Szekeres G (1935) A combinatorial problem in geometry. Compos Math 2:463–470
4. Mysak L, Szekeres G (1966) Behavior of the Schwarzschild singularity in superimposed gravitational fields. Can J Phys 44:617–627

Chapter 6
My Harvard Teaching Assistantship: Life as an Academic Doppelgänger

Abstract To help cover my tuition and living costs at Harvard, I secured a teaching assistant (TA) position in the Harvard engineering school where I was enrolled to do a PhD in applied mathematics. Since I had done an honors course in the theory of complex variables as an undergraduate, I was going to be a TA in this subject. However, when tested on some complex variable problems by the chair of the applied mathematics program, he concluded that I should take this course for credit because of its more applied orientation. Did I lose the TA? No, the chair did not know of any rules at Harvard that prevented me from both taking the course for credit and being a TA for it.

After I arrived in Cambridge, Massachusetts, in September 1963 to begin a PhD in applied mathematics at Harvard, I was illegally camping out in an empty dorm room a friend found for me. It was hot and sticky compared to the cool and dry Canadian prairies I left behind. My top priority was to get on a waiting list for a permanent room in a Harvard graduate dormitory. Since the dorm application would take a few days to process, I went over to Pierce Hall to find Professor George Carrier, the chair of the applied mathematics program. I wanted to start selecting my courses for the fall semester.

Professor Carrier was in a short-sleeved shirt trying to keep cool in front of a fan. "Welcome to Harvard," he said while leaning back in his swivel chair, with one foot tucked under his bottom. "It's been a while since we've had a Canadian graduate student in applied mathematics here, but I've heard you have strong honors programs at your universities."

"I think we do," I replied and then described the advanced courses in mathematics and physics I had taken at the University of Alberta. "One of my favorites was the theory of complex variables." Little did I know what this was leading to.

Professor Carrier then sprung out of his chair toward the blackboard and quickly drew a picture of a circle with a diving board resting on the top. "How would you map the exterior of this object onto the upper half plane in x–y space?" he asked.

L. Mysak, *Adventures in Climate Science, Ocean Waves, and the Flute*, Springer Biographies, https://doi.org/10.1007/978-3-032-19848-8_6

I gulped. I knew from my complex variables course that this was theoretically possible, but I had no experience in how to use the Schwarz–Christoffel transformation that is needed to solve this problem. Failed question number (1) Carrier next asked me to evaluate a contour integral involving branch point singularities in the complex plane. I could evaluate such an integral with pole singularities, but not one with branch point singularities. Again, I was stymied; failed question number (2) Then he asked me about the method of steepest descent. I had no clue what he was talking about.

"Well," he said, "I think you should take our basic graduate course in complex variables. This focuses on solving the sort of problems I have just asked you."

I started sweating. To help cover my tuition and living expenses at Harvard for the year, I was offered a Teaching Assistant (TA) position for this very course. When I pointed this out to Carrier, he looked at the ceiling, paused for a minute, and then concluded, "I don't think there are any rules at Harvard that prevent you from being both a TA and a student in the same course."

Whew, I said to myself, and then thanked him for this rather unusual news. This was, in fact, very good news since the stipend for the TA would cover about 20% of my annual expenses and a Harvard scholarship would pay for another 60%. Since I had taught summer school at the University of Alberta before coming to Harvard, I was able to cover the rest of the expenses from my savings.

More good news came 2 days later: I got a double room in the graduate dorm Perkins Hall. My roommate was economist Dick Freeman, a graduate of Dartmouth College; I soon learned that he was very ambitious. Dick worked 16 hours a day and alternately took wake-up and sleeping pills. Two decades later he became a distinguished Harvard professor.

To my surprise, Carrier taught the complex variables course, which had over 150 students. With a class of this size, four TAs were needed to grade the bi-weekly assignments. However, these grades didn't count toward the final mark. In most courses at Harvard, this grade was based solely on the final exam. No pressure for a Canadian, eh?

I enjoyed collaborating with the other TAs, first to work out the solutions to the assigned problems, and then to divvy up the marking responsibilities. I certainly learned a lot in the process about the applications of complex variable theory to real-world situations in engineering science and physics. My biggest challenge as a TA was to not let on to several of my new dorm friends who were taking the complex variables course that I was also grading their assignments. I managed to keep this secret until 2 weeks before the final. At that point Mike Ioffredo, a dorm friend doing a PhD in physics, happened to see a pile of complex variable assignments on my desk. "Oh, so you're the one who's been giving us the extra tips on how to solve the complex variable problems!".

Mike was an exceptional student. He got 100% on the final exam, which Professor Carrier marked. Even with my experience as a TA and thorough preparation for the final, I only managed 90%. But anyone who got 85% or more on the final was assigned an A for the course. This made us both happy.

Professor Carrier was a wizard in the course lectures. He came to the classroom without any notes and worked through every theorem and application starting from basic principles. I finally learned how to do integrals with branch points and apply the method of steepest descent, both of which I would soon use in my research. Twenty years later, I visited Ioffredo and his young family in Washington, DC, where he now worked in the Pentagon. He was doing operations research, a field far removed from his PhD studies in particle physics. When I told him about the seminar on waves that I gave at the US Office of Naval Research which involved the method of steepest descent [1], Mike's face went blank. He couldn't remember learning about this method in the complex variables course we took together. Maybe I wasn't such a good TA after all.

I also took an inspiring fluid mechanics course from Professor Carrier that fall. The topics covered in the course really excited me, especially those about waves. The following year I asked Carrier if he would supervise my PhD thesis. However, he was not available and suggested that I might consider oceanographer Allan Robinson, a new faculty member, as a potential supervisor.

I took his advice to heart and ended up doing a PhD thesis on ocean waves under Robinson's guidance (Chap. 7). It was a good choice because I was able to complete the PhD program in 3 years (a rare event in those days). And thanks to Robinson's good reference, I later secured a tenure-track position at UBC in applied mathematics and oceanography. My unusual interactions with Robinson are described in Chap. 7.

Acknowledgements I thank Janet Boeckh and Josef Schmidt for their insightful comments.

Reference

1. Mysak LA (1983) Generation of annual Rossby waves in the North Pacific. J Phys Oceanogr 13:1908–1923

Chapter 7
Surviving My Harvard PhD Supervisor: Working Through the Graduate Student Labyrinth

Abstract I was lucky to complete my PhD at Harvard in 3 years. When I started my studies there in 1963, I heard graduate students describing themselves as G7s or G8s. Even after 7 or 8 years at Harvard, they still hadn't submitted their theses. My supervisor, Allan Robinson, is part of the reason why I took just 3 years. He was very supportive when I struggled with my PhD oral qualifying exam, and he suggested a couple of quite tractable problems to work on for my thesis. His style of supervision was very much hands off at the beginning, which suited me just fine since I already had experience doing research as an undergraduate and as a master's student in Australia. I was able to make good progress in my research after about 1 year, and Robinson then suggested that I write up a draft of my thesis and show it to Professor George Carrier, the chair of the applied mathematics program. Carrier was impressed with the draft and immediately suggested that I could schedule a thesis defense in September 1966, before Robinson went on sabbatical.

7.1 Embarking at Harvard

During my first year at Harvard in the PhD applied mathematics program, I took two courses from Professor George Carrier, whom I admired for his brilliance and knowledge of oceanic dynamics. He was my first choice for a PhD supervisor. When I asked him in September 1964 to consider taking me on as a PhD student, I got a disappointing response.

"Larry (as I was then called), normally I would be delighted to do this, but next year I'll be away on sabbatical," he replied. "Why don't you consider Allan Robinson as a potential supervisor? He's a new faculty member on the lookout for motivated graduate students."

At the beginning of my second year at Harvard, I took the course "Geophysical Fluid Dynamics" (GFD), which explored the large-scale circulations in the atmosphere and ocean. I was fascinated by fluid motions and wanted to do a PhD thesis research in this area. This course would certainly give me a good background to carrying out this type of research. Allan Robinson was the instructor for the course.

L. Mysak, *Adventures in Climate Science, Ocean Waves, and the Flute*, Springer Biographies, https://doi.org/10.1007/978-3-032-19848-8_7

Robinson's GFD lectures were pedantic but informative. I quickly learned about the effects of the earth's rotation on large-scale currents in the atmosphere and ocean, and how the ocean temperature and salinity affected the ocean's density and overturning motion. I got annoyed at the snarky interjections by Jim Baker and Peter Niiler, Robinson's postdoctoral students who sat at the back of the class. However, this activity livened up the classes and introduced me to the give-and-take of academic debate. Later both Jim and Peter helped me with the last chapter of my PhD thesis.

There was no final exam for the course; instead, each student had to write a 10-page term paper about some topic in oceanic or atmospheric dynamics. Toward the end of the term, I went to Robinson's office to discuss possible topics for a paper on ocean dynamics.

"What is your interest in the oceans?" Robinson asked.

"I'd love to write about ocean waves," I replied.

I was always intrigued by wave motions, and in Professor Carrier's fluid mechanics course that I took last year, the discussion about waves in fluids made me eager to learn more. Robinson then pulled out of a filing cabinet a short paper he recently published [1]. It was entitled "*Continental shelf waves and the sea level response of the sea surface to weather systems*."

"Take a look at this," he said. "For your term paper I want you to explain the physics and expand on the mathematical operations in this paper. Also, you should describe the observational evidence for these waves."

When I dug into this paper, I discovered that the term "continental shelf waves" was newly coined by Robinson. Through an elegant analysis of the shallow water equations for fluid motion on a continental shelf, he came up with a theory of coastal trapped (or "shore hugging") waves traveling along a sloping shelf (hence the term continental shelf waves). This wave theory could explain the northward traveling sea level signals with periods of a few days that were recently observed along the east coast of Australia [2, 3]. Since there were no other published papers on shelf waves except these three, I wouldn't have to do an extensive literature review for my term paper. Thus, it was brief and good enough to get an A for the course.

7.2 Setting Course for My PhD Thesis

During the spring semester, I took a one-on-one reading course from Robinson in which I attempted to generalize his theory for the case of shelf waves traveling around a large circular continent that crudely modeled Australia. As I seemed to be making good progress on this project, toward the end of the semester Robinson pulled me into his office one day and said, "I want you to be one of my PhD students."

I was stunned. Usually, it was the other way around at Harvard. During your second year you were expected to visit various professors and beg one of them to take you on as a research student. However, even though I now had a supervisor, I had to pass a PhD oral qualifying exam in June 1965.

I'll never forget that oral exam. The examiners represented half a dozen different areas of applied mathematics and engineering science, and their challenging questions forced me to "tread in the water" as I tried to work out the answers in my head. I managed to answer two of them well, but I stumbled through most of the others. Robinson took pity on me near the end of the two-hour examination period and said, "Pose your own question in GFD and give us the answer."

I tried to explain the phenomenon of baroclinic instability of large-scale flows in the atmosphere, a process which leads to the high- and low-pressure weather systems at mid-latitudes. However, it was evident to the examiners that I had a poor understanding of all the mechanisms involved. To save the day, Robinson persuaded the examining committee to give me a conditional pass for my oral: During the next week I had to write a report to clearly explain the nature of baroclinic instability and its role in both oceanic and atmospheric dynamics. Fortunately, the report was accepted, and I was free to formally start my thesis research under the direction of Allan Robinson. Whew, I could start sailing again.

Reflecting today on my miserable PhD qualifying exam six decades later, I think there is much merit in what DelSole and Dirmeyer [4] have introduced in the Department of Atmospheric, Oceanic, and Earth Sciences at George Mason University to replace this traditional qualifying exam. Rather than a short oral exam which most students find very stressful, they have introduced a one-semester qualifying process in the second year of a PhD program which is aimed at testing the student's ability to do research. This process evaluates a student's ability to formulate a research question, explain it to a group of professors, carry out independent research, and then write a paper for journal publication.

I soon discovered that Robinson used the "puppy method" of supervision when starting off his new PhD students. He expected us to work on our own and struggle to find our own solutions, like the young puppy tossed into the middle of a pond. If the puppy makes it to shore by itself, great. If the puppy starts to flounder, however, a lifeline is thrown out.

At our first meeting, Robinson stated "I want you to complete your mathematical solution for shelf waves around a large circular continent, a project that you started in the reading course this past spring." He continued with "Then try to find a mathematical formula for the sea level energy spectrum that would characterize the response of the ocean surface to an observed atmospheric weather system traveling eastward across Australia."

He didn't set a time for our next meeting. As a Robinson student, I quickly learned that if I were to solve the problems posed mainly by myself, I would soon be out of Harvard with my PhD. This suited me fine because later that summer I was planning to get married and wanted to start looking for a permanent position. On the other hand, if I needed a lot of help with my research, I could be a G7 or G8 student grinding through my thesis research and living in semi-poverty as a married student.

That summer I made some progress in my research without his help, but I was preoccupied with my upcoming marriage to Diana in Edmonton followed by a honeymoon in the majestic mountains of Jasper Park, Alberta. When we returned to Boston in September, I was refreshed and ready to plod on with my thesis research. After a month, however, I was really stuck and made an appointment to see Robinson. I filled his blackboard with many equations which I hoped would lead to the solution of the island shelf wave problem. "What should I do next?" I asked in desperation.

Robinson listened carefully as I made my presentation, but at the end he had nothing to offer in the way of solving my mathematical problem. After a long pause he replied, "Larry, if one method doesn't work, you'll just have to find another."

Fortunately, a couple of months later, I did find a solution to the island shelf wave problem after some helpful discussions with my fellow PhD students. Next, I was able to determine theoretically the effects of offshore stratification and a longshore current on the shelf wave speed. It was now spring 1966. To complete the thesis, I still needed to find a mathematical expression for the shelf wave energy spectrum and compare this with the observed spectrum. As Robinson was going to England for a sabbatical in September 1966, I needed to get this done quickly. In June his postdocs Jim and Peter came to my rescue. We met together in a small classroom and brainstormed with Robinson on how to use the methods of time series analysis for linear systems to solve this spectral problem. We quickly came up with the solution, and I was now able to put together a draft of my thesis.

In July 1966 I gave the draft to Robinson. He looked at it briefly and then suggested I send a copy to Professor Carrier, who was on a holiday in the remote woods of Maine. Clearly, Robinson wanted the approval of the chair of the applied mathematics program before he allowed me to submit my thesis. After about three weeks, I phoned Carrier for his comments. "The thesis is short, but the results are new and significant," he said. He then told me that I could now set a date for my PhD thesis defense. Yahoo!

I successfully defended my thesis [5] in early September 1966, and a week later I passed my language requirement for the PhD. I did Russian for this since I had taken two courses in Scientific Russian as an undergraduate at the University of Alberta. This knowledge of Russian proved to be very useful later when I was traveling by car through Russia and Ukraine in the early 1970s during the height of the cold war (Chap. 32).

Indeed, I was lucky to have completed all the requirements for the PhD in just under 3 years. Because I was married, I was highly motivated to finish my graduate studies as quickly as possible. In addition, my previous experience in research and writing up the results, first as a summer student working for the Canadian Defense Research Board and later as an MSc student doing research in Australia, helped a lot. Above all, Robinson had given me what turned out to be an "easy problem" for my thesis. Other Robinson PhD students often took several years to complete their theses. I later heard that a few of his students were mentally scarred by their thesis research experience at Harvard.

7.3 New Voyages

My next major encounter with Allan Robinson was in Cambridge, England, in spring 1972. We both happened to be on sabbatical as "senior visitors" in the Department of Applied Mathematics and Theoretical Physics (DAMTP), University of Cambridge. After finishing my PhD and spending a brief postdoctoral period at Harvard to write up my thesis for publication [6, 7], I started my academic career as an assistant professor of mathematics at Vancouver's University of British Columbia (UBC) in September 1967 (Chap. 10). After 4 years at UBC, I was eligible for a sabbatical and chose DAMTP because of the many researchers and visitors there working in GFD. I had had little contact with Robinson once I left Harvard. I think this is because I often felt intimidated by his sophisticated New England upbringing and Ivy League arrogance. Also, our research interests went in different directions. I continued to work on shelf waves and related coastal problems for the next decade [8], whereas Robinson focused on mid-ocean eddies and the Gulf Stream [9]. Thus, it came as a surprise when one day in Cambridge he suggested that we have dinner together at his favorite country inn.

We drove to the inn in his Alfa Romeo, a car he purchased overseas to bring back to the USA. I remembered that he always had a taste for fancy sports cars. The inn had a beautiful garden with peacocks strutting about, and we were able to eat on the terrace. The food was excellent. I noticed that Robinson chose the most expensive meal on the menu. He also ordered a cigar to smoke with his coffee. Our conversation was relaxed as we focused mainly on trying to understand the lifestyle of British academics. I recall that the bill came to nine pounds. Allan pulled out a fiver, and I happened to have four one-pound notes sticking out of my wallet, which he noticed.

"That's fine for your share of the bill," he stated. "Let's keep the professor-student relationship intact."

"Would you like to come to my flat for a brandy when we get back to town?" I offered.

"Great idea," he replied. "I want to talk to you about some serious scientific issues."

What the hell does this mean, I thought to myself. Time would tell.

After the first sniff of brandy, Robinson came right out with a question I had been struggling with as an applied mathematician in a large department of mostly pure mathematicians. Did I belong there, or would I be better off in an oceanography department since I was using mathematics to solve various ocean wave problems?

"Larry," Allan asked, "are you more interested in working as a methodologist or as a phenomenologist?" By this he meant that a methodologist has a bag of mathematical tools and then looks for problems in the physical world to apply them. A phenomenologist, on the other hand, is more interested in an observed feature in nature which he or she is trying to understand; the mathematical tools come into play later as needed to rigorously describe the feature.

“I’m not sure,” I replied. “I have probably been thinking of myself mainly as a methodologist. But perhaps I should think more about the other approach to my research.”

When I returned to UBC in fall 1972, I thought again about the question that Robinson posed. After doing research with a few observational oceanographers at UBC and elsewhere, I discovered that it was indeed thrilling to be able to use my mathematical tools to model and explain a variety of puzzling oceanic and climatic observations. Thus, the dinner with my former PhD supervisor proved to be a game changer for me. Over the years, I have studied, for example, the impact of natural climate variability in the Northeast Pacific on sockeye salmon migration, the influence of sea ice cover variations in the Greenland Sea on the overturning ocean circulation in the northern North Atlantic, the effects of vegetation and the Earth’s orbital variations on glacial inception, and the impact of global warming on outdoor skating rinks (see Chap. 24 for more details).

7.4 Final Log Entries

The last time I met Allan Richard Robinson (known by his colleagues as AR^2) was in December 2005 when I happened to be giving a seminar on climate dynamics at MIT, in Cambridge, MA. After the seminar, he invited me to his office at Harvard. Robinson looked happy in his three-piece suit, and he enthusiastically described his new work on the theory of physical-biological interactions in the ocean. He then led me to a corner shelf which displayed the theses of his nearly 30 PhD students. I was his fourth student, and it was clear that he was proud of his academic sons and daughters.

Recently I discovered that I was one of eight students Robinson chose to list in his academic family tree (Fig. 7.1). I knew that his PhD supervisor, his academic father, was Professor Carrier, but from this tree I also learned that my academic great-great-great-grandfather is the eminent German fluid dynamics professor Ludwig Prandtl. He was the first one to describe the boundary layer in flows with very little friction, such as those around an airplane wing. The tree also shows that Stephen Timoshenko is my great-great-grandfather. He is widely regarded as the “father of applied mechanics” in the USA. Timoshenko was born in Shepetivka, Ukraine and did graduate studies at the University of Göttingen under Prandtl. Before coming to the USA, he was a professor at the Kiev (now Kyiv) Polytechnic Institute. Being of Ukrainian descent, I want to pass on this discovery to my children, Paul and Claire.

Allan Robinson died unexpectedly of cardiac arrest in 2009. He was 76 but still active in research. While I was never close to him on a personal level, I greatly appreciate the support that he gave me as a young PhD student six decades ago. During my 50 years of active graduate student supervision (1967–2017), I have tried to offer similar support to my students. However, I never used the puppy method when my students started out on their thesis projects.

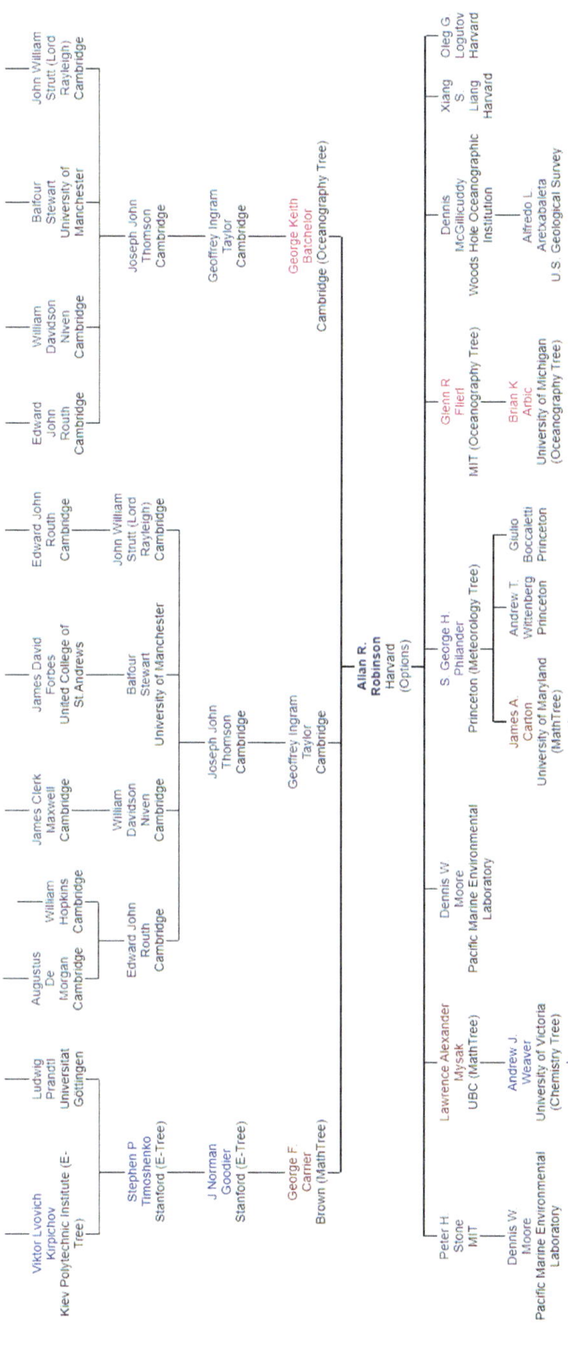

Fig. 7.1 The academic family tree of Allan Robinson. The eight academic children of Robinson shown in the tree include Peter Stone (PhD 1964) and three students doing research in the mid to late 1960s: Lawrence Mysak (PhD completed in 1966 but conferred in March 1967), Dennis Moore (PhD 1968) and George Philander (PhD 1970). From Physics Tree—Allan R. Robinson Family Tree (academictree.org)

Acknowledgements It is a pleasure to thank Janet Boeckh for her comments on an early version of this memoir, Josef Schmidt for his refined literary critique, and Tom Beer for his insightful suggestions.

References

1. Robinson AR (1964) Continental shelf waves and the sea level response of the sea surface to weather systems. J Geophys Res 69:367–368
2. Hamon BV (1962) The spectrums of mean sea level at Sydney, Coff's Harbour, and Lord Howe Island. J Geophys Res 67:5147–5155
3. Hamon BV (1963) Correction to "The spectrums of mean sea level at Sydney, Coff's Harbour, and Lord Howe Island." J Geophys Res 68:5635
4. DelSole T, Dirmeyer P (2024) Fixing the PhD qualifying exam. Phys Today 77:34–40
5. Mysak LA (1966) Continental shelf waves. PhD Thesis, Harvard University, pp 69
6. Mysak LA (1967) On the theory of continental shelf waves. J Marine Res 25:205–227
7. Mysak LA (1967) On the very low frequency spectrum of the sea level on a continental shelf. J Geophys Res 72:3043–3047
8. Mysak LA (1980) Recent advances in shelf wave dynamics. Rev Geophys Space Phys 18:211–241
9. Robinson AR, Harrison DE, Haidvogel DB (1979) Mesoscale eddies and general ocean circulation models. Dyn Atmos Oceans 3:143–180

Chapter 8
Memories of a Conference in Moscow

Abstract During my final year as a PhD student at Harvard in 1966, my supervisor arranged for me to attend an international oceanography conference in Moscow. During my two-week time there, I met Russian graduate students as well as internationally renowned oceanographers. As this was the height of the cold war, I felt certain that I was always under constant surveillance by the KGB (the Russian secret police). And one day I had a frightening experience when two well-dressed strangers joined me for a coffee in a large park and started asking me many questions about America.

While at the conference I shared a hotel room with Robert Deitz, who coined the term "sea floor spreading". At one of the conference sessions, I heard Canadian oceanographer Robert Stewart OC FRS FRSC deliver an inspiring lecture on ocean turbulence. A year later, he helped me get a tenure track position at UBC, where I would spend the first two decades of my academic career as a professor of mathematics and oceanography.

In June 1966, I flew to Moscow to attend the Second World Oceanography Conference. This was just a few years after the Cuban missile crisis, and we were at the height of the Cold War. While I was excited to be making this trip at the young age of 26, I was also apprehensive. Colleagues warned me to expect a debriefing from the CIA when I returned to Boston.

I had nearly finished my PhD thesis at Harvard on continental shelf waves [1], a topic dear to the heart of my supervisor, Allan Robinson. Since I was then reasonably fluent in Russian and a highly motivated student, Robinson thought I would benefit by attending this international meeting, my first, where I would meet oceanographers from all corners of the world. And indeed, I did.

All my expenses for this two-week conference were paid, thanks to a special travel fund that Robinson held in trust. When I arrived at the Hotel Ukraine in Moscow, I was assigned a room to be shared with Dr. Robert (Bob) Dietz, a 51-year-old marine geophysicist who worked for the US Coast and Geodetic Survey. In the early 1950s Bob Dietz discovered fracture zones in the Pacific Ocean floor, and in 1961 he coined the term "seafloor spreading" [2] to describe the movement of the seafloor away

L. Mysak, *Adventures in Climate Science, Ocean Waves, and the Flute*, Springer Biographies, https://doi.org/10.1007/978-3-032-19848-8_8

from the mid-Atlantic Ridge, which was formed by rising basalt from the earth's interior. Only recently did I learn [3] that Dietz's contributions to our understanding of continental drift are considered world class.

Back in 1966, Dietz and I didn't have much in common scientifically. However, I do remember his loud snoring at night, which I thought was most appropriate for one who studied plate tectonics and shaking movements of the seafloor. When I told him that his snoring kept me awake at night, Bob said, "try using these", as he put a pair of earplugs in my hand. But I also wished that I had his eye shades, since the bedroom was already light by 4 am, another reason making it difficult to get a good night's sleep.

One morning before the conference started, I explored the vast campus of Lomonosov University, where the scientific papers would be presented. As the hotel breakfast was meager, I was soon very hungry; luckily, I found a university cafeteria for lunch. Since the tables were all taken, I joined a large round one with several dark-haired, black-eyed students who came from the "stan" republics (Kazakhstan, Uzbekistan, and Kyrgyzstan). Then I introduced myself in Russian. They were amazed.

"Where did you learn Russian?" they asked. "And why are you here?"

"At the University of Alberta," I answered for the first question. "After the launch of Sputnik 1 in the 1957," I continued, "scientific Russian classes were introduced and many of us in science programs took them." This was my first use of Russian outside the classroom.

We subsequently had a lively conversation about student life in Moscow far from home. After finishing their degrees (most were studying engineering), these students wanted to return to their countries as soon as possible. They missed their families and wanted to help develop local industries.

On my second day in Moscow, while I was waiting in the hotel lobby for transport to the conference site, a young, friendly man approached me and said (in Russian-accented broken English), "I look for Professor Robinson and student. I am student in Institute of Oceanology in Moscow, and I be guide."

"That's quite a coincidence," I replied. "I think I'm the student you're looking for, and my name is Larry Mysak." (When living in Australia during 1962, I took on the name Larry, and I did not revert to Lawrence until 1972.)

"My name is Boris Gavrilin, student of Professor Monin of USSR Academy of Sciences. I study ocean turbulence."

We did not see Professor Robinson around the hotel lobby that morning, so the two of us went off to the conference in the prearranged bus. Boris spoke English reasonably well, but it was easier for us to converse in Russian. Thanks to Boris, we quickly found the correct rooms for the conference lectures and then spent most of the day together. We talked little about our research. Boris was more interested in my life as a student in America and what I did outside the classroom. He especially asked whether I had any novels in English that I could give him!

During the conference period Boris took me to several touristic sites in Moscow (such as the tomb of Lenin), and to science museums and the space center. The Russians are very proud of their accomplishments in rocketry and space travel, then in its infancy. Boris also wanted to introduce me to his supervisor, and he tried to give him a draft of a paper he was writing. However, Professor Monin was always busy talking to the bigwigs at the conference and kept giving Boris and me the brushoff. This was very frustrating for Boris, but not an uncommon behavior of many top academicians in the USSR. I felt sorry for Boris.

We exchanged a couple of letters after I returned to North America, but I did not expect to meet him again. To my surprise, when I was at an International Association for the Physical Sciences of the Oceans (IAPSO) conference in Melbourne in January 1974 (see Chap. 22), he and many of his colleagues showed up on a Russian oceanographic research vessel that sailed from Vladivostok. They couldn't afford to stay in hotels but were happy to sleep on the ship and come to shore for conference lectures. A few of us were invited on board their ship for vodka and sausage, and my friendship with Boris was renewed.

One day I heard Robert (Bob) Stewart OC FRS FRSC, one of Canada's leading oceanographers from UBC, give a talk on ocean turbulence measurements off the BC coast. This was of great interest to the Russian navy. While I did not understand the significance of his work, the results were clearly and confidently presented, with a few Russian phrases thrown in. The response of the local oceanographers was very positive. I later met Professor Stewart in our hotel, and he encouraged me to return to Canada after my studies at Harvard. He was impressed that I spoke Russian. In the fall of 1966, while I was doing a postdoc at Harvard, I was offered a tenure-track position in UBC's Department of Mathematics and associate membership in the Institute of Oceanography, which was for graduate studies only (Chap. 10). I'm sure that Stewart must have put in a good word to the UBC administration for me to get this offer.

Several times I shared a dinner table with my American colleagues at restaurants reserved for foreigners. The menus were often only in Russian, and the selection was limited. I tried my best to translate the menus for the others, but what arrived at the table (after a long wait) was sometimes unrecognizable. Later, just one bill came for the whole table. Since it was hard to make out its details, I naively divided it up evenly according to the number sitting at the table. However, more than once I thought that I was paying too much for the simple food I ordered (perogies and kielbasa). The professors at the table, on the other hand, always seemed to be having steaks and wine and thus could have offered to pay more.

Most of the time I felt safe in Moscow when walking around the center of the city. But one day a few of us strolled over to an outdoor café in a park, about two kilometers from our hotel. It was warm and sunny, and we felt very relaxed until we were joined by two smartly dressed gentlemen who were not from the conference. They said they were journalists, and when I asked them repeatedly to explain what kind of work they did, I always got the reply "rabot" (the Russian word for work). But when they started asking us detailed questions about political life in America, we became suspicious and worried. Were they from the KGB? Two of my colleagues got frightened and quickly left. I persuaded one friend to stick with me until I finished my coffee and cake, but I too was happy to return to the hotel, especially when these strangers invited us to their apartment. I was shaking when I got back to my room.

I left Moscow on an Aeroflot flight to London; I was happy to get out of Moscow safely without any other scary incidents. On this flight I sat at a table with two seats on either side, a configuration commonly found on trains. Opposite me was Englishman Dr. Michael Longuet-Higgins FRS, who was also attending the conference. Longuet-Higgins was a world leader in ocean wave research, but he came across as being very reserved. However, when I told him about my thesis work concerning shelf waves travelling around a large continent that modelled Australia, he got very excited and quickly pulled a large folder out of his brief case to show me his latest calculations on the theory of long waves trapped around seamounts. Our research interests were surprisingly close, and we later exchanged reprints of our work. A year later, Longuet-Higgins moved to Oregon State University in Corvallis, and to my delight, he invited me to spend the summer of 1968 working there as his research associate. Longuet-Higgins returned to England shortly afterward as a Royal Society Professor at the University of Cambridge, and he was my academic host when I spent a sabbatical at Cambridge in 1971–72 (Chap. 11). The moral of the story—be friendly to your seat partner on an airplane!

The next and last time I visited Moscow was in August 1972, when I gave a paper on wave propagation in random media at the International Union of Theoretical and Applied Mechanics (IUTAM) Congress. I now had tenure at UBC and felt confident about making a presentation to an international audience. This was still during the Cold War, but I had no frightening experiences at this conference. However, I was somewhat intimidated when at the end of my presentation, a middle-aged Russian scientist Yuri Miropolsky asked me to come with him to meet the eminent Russian Academician Brekhovskikh [4] who was working on waves in layered media. They were puzzled by the strange propagation properties resulting from my theoretical calculations on internal waves in a randomly stratified fluid and wanted a more detailed explanation. I couldn't do this at the time, and later I discovered why. In 1974, my postdoc Charles Tang found a mathematical error in the paper I presented in Moscow, and when it was corrected, the strange properties disappeared [5].

Three decades after meeting Bob Dietz in Moscow, I was lecturing on continental drift in an introductory oceanography course at McGill. At the same time, I was doing research on the global thermohaline circulation (THC) and its role in warming the polar regions through the transport of heat from the tropics to the higher latitudes. This global circulation pattern is also known as the "conveyor belt" [6]. The part of this circulation that is in the Atlantic Ocean is known as the Atlantic Meridional Overturning Circulation (AMOC). One day after class, I asked my postdoc Gavin Schmidt to think about modeling the THC when there was no Atlantic Ocean, the case about 100 million years ago before seafloor spreading, as described by Dietz [2]. Using a simple zonally averaged model of the ocean circulation developed at McGill, we were able to show [7] that there was a THC pattern in a large proto-Pacific Ocean that could warm the polar regions to an extent that is consistent with ancient ocean data collected from deep ocean sediment cores. Unfortunately, Dietz died in 1995, and I could not share these theoretical results with my 1966 roommate in Moscow.

8.1 Coda

This memoir was written a year after Russia's brutal invasion of Ukraine in February 2022. Since I am of Ukrainian ancestry, my view of Russian scientists is now complicated. When I was president of IAPSO (2007–2011), one of the vice-presidents was Eugene Morozov, an oceanographer from Moscow. Eugene went on to become president of IAPSO (2011–2015), and we had a very cordial relationship for many years. However, when Montreal hosted an IAPSO assembly in 2019, many Russian oceanographers were denied visas to attend this meeting, probably because of the annexation of Crimea, by Russia in 2014. Crimea had been part of Ukraine since the late 1950s. Many collaborative projects with Russian scientists were canceled, and now I have no contact with them. Such are the impacts of war and politics. Even after the war ends, it will take many, many years to build trust and friendships with Russian scientists.

On an endnote, after the 1966 trip to Moscow I did end up with a lasting memento that I did not expect—a dog! While I never was debriefed by the CIA when I returned to Boston, my then wife, Diana, decided that we needed a pet to keep her company when I was away at my next two-week conference. That summer we thus acquired Jasper, a large black poodle, who gave us a lot of joy.

Acknowledgements I thank Janet Boeckh for her comments on the first draft of this memoir, and Josef Schmidt and Tom Beer for their insightful suggestions for improvement of a later version.

References

1. Robinson AR (1964) Continental shelf waves and the sea level response of the sea surface to weather systems. J Geophys Res 69:367–368
2. Dietz RS (1961) Continent and ocean basin evolution by spreading of the sea floor. Nature 190:854–857
3. Oreskes N (2021) Science on a mission: How military funding shaped what we do and don't know about the ocean. University of Chicago Press, Chicago
4. Brekhovskikh LM (1960) Waves in layered media. Academic Press, New York
5. Tang CL, Mysak LA (1976) A note on 'Internal waves in a randomly stratified fluid.' J Phys Oceanogr 6:243–246
6. Broecker WS (1991) The great ocean conveyor. Oceanography 4:79–90
7. Schmidt GA, Mysak LA (1996) Can increased poleward oceanic heat flux explain the warm cretaceous climate? Paleoceanography 11:579–593

Chapter 9
The Invited Talk

Abstract In September 1966, I gave my first invited talk at an international symposium on ocean waves in Washington, DC. I was there as a substitute for Harvard Professor George Carrier, Member NAS, who was invited to speak about tsunamis. Carrier was bored with the thought of having to give this talk again, and at the last minute he asked me, as a newly minted PhD, to go to Washington in his place. He encouraged me to talk about my thesis research on continental shelf waves. What an honor! But it was a frightening experience since all the other invited speakers were twice my age. I survived however, and now, looking back six decades later, I see how important it is for the career of young scholars to give invited papers. Just do it!

A few days after I defended my PhD thesis [1] at Harvard in September 1966, Professor George Carrier, chair of the examining committee at the defense, called me into his office. "Lawrence, I have a favor to ask you," he said. "At the end of September, I am scheduled to give an invited talk on tsunami-generated ocean waves at the *Sixth Symposium on Naval Hydrodynamics* in Washington, DC. Many colleagues have heard this talk, and I am bored giving it. I'd be grateful if you went in my place to speak about your thesis work. Your expenses will be paid."

I gulped and felt stressed. I was just 26. As a very newly minted PhD, I knew I'd be nervous speaking to ocean wave experts who would be attending and presenting papers at this conference.

"Don't worry," Professor Carrier assured me. "You explain things well, and the oceanography community needs to hear about these mysterious continental shelf waves."

Today, I appreciate what a big step this was so early in my career. When serving on appointment and prize committees, I search the candidates' CVs for the list of invited talks. Where have these been given? How often? Are the talks of an overview nature, or do they deal with cutting-edge science? For scientists early in their career, it would be unusual to see a long list of invited talks.

While I don't remember how my talk went that September, I do recall having several questions afterward. And I admit that I had trouble answering some of them. I naively thought the dark blue suit that I wore for the occasion would help smooth

L. Mysak, *Adventures in Climate Science, Ocean Waves, and the Flute*, Springer Biographies, https://doi.org/10.1007/978-3-032-19848-8_9

over these challenges. It did not—my knees trembled during the question period as I searched for answers. Now, many talks later, I still get nervous when the question period starts.

However, I do remember well some of the other invited talks given by established oceanographers, most of whom were twice my age. Klaus Hasselmann, from Germany, impressed me the most. (Fifty-five years later, he was a co-recipient of the Nobel Prize in Physics in 2021.) In 1966 he published a review paper on the application of Feynman diagrams to describe nonlinear wave-wave interactions in the ocean [2]. I found it remarkable that he was able to use the methods of theoretical physics developed by Nobel Laureate Richard Feynman to represent how energy is transferred between wave groups propagating across the ocean in different directions. He talked like a wizard, and he mesmerized us all with his presentation of very, very long equations that he used to calculate the amount of energy transferred between these wave groups. I didn't follow his talk very well at the time, but 10 years later, I had to become familiar with his pioneering work when my colleague Paul LeBlond and I were writing a chapter about wave-wave interactions for our book *Waves in the Ocean* [3].

Later that fall, I submitted a paper based on my PhD thesis to the Journal of Marine Research [4]. I was not amused when an anonymous reviewer raised a question that was also asked at the Naval Symposium. Clearly, this reviewer heard my talk in Washington that September and was still waiting for a better answer to his question. Over the years, I have experienced this type of interaction with my colleagues several times. So, it's best to be polite and respectful when on stage as an invited speaker.

Most of the international figures who were at that Naval Symposium have passed away by now. But one who had a very long life was Walter Munk, Member NAS, from La Jolla, California. He celebrated his 100th birthday in summer 2018 and sadly passed away in February 2019 at age 101. Walter was an expert on internal waves, tides and mixing in the ocean, and his ideas on these topics are still highly respected by the oceanographic community. In 2001 he was the inaugural recipient of the Prince Albert I Medal, which was established by the International Association for the Physical Sciences of the Oceans (IAPSO) to recognize biennially a prominent scientist in this field.

Walter's interest in the work of other colleagues during his lifetime is legendary. After receiving this medal at the 2001 IAPSO meeting in Mar del Plata, Argentina, he joined a group of us in the hotel bar. After a few pleasantries, he asked, "Lawrence, are you still working on continental shelf waves?" What a memory!

During my career as a professor and supervisor for scores of graduate and post-doctoral students (Chap. 24), I have relentlessly encouraged them to give papers at conferences and other universities, especially if they are invited to do so. Most are glad to go out into the world to present their work and make contacts with other colleagues in their fields. But occasionally there is some resistance. One recent PhD student said, "Do you realize that you're taking me outside of my comfort zone?".

"Really?" I replied. "If you want to be a successful scientist or professor one day, then this is the norm. The sooner you get used to this, the better."

Acknowledgements I thank Janet Boeckh for her helpful comments on a first draft of this chapter.

References

1. Mysak LA (1966) Continental shelf waves. PhD Thesis, Harvard University, pp 69
2. Hasselmann K (1966) Feynman diagrams and interaction rules of wave-wave scattering processes. Rev Geophys 4:1–32
3. LeBlond PH, Mysak LA (1978) Waves in the Ocean. Elsevier Scientific Publishing Co, Amsterdam
4. Mysak LA (1967) On the theory of continental shelf waves. J Marine Res 25:205–227

Chapter 10
My First Academic Home: University of British Columbia

Abstract In September 1967, I became an assistant professor of mathematics at the University of British Columbia (UBC) in Vancouver. Although my main assignment would be to teach applied mathematics courses for science and engineering students, I looked forward to collaborating with researchers at UBC's famed Institute of Oceanography since my PhD thesis was on the topic of ocean waves. After joining the UBC faculty, I also learned that there were many faculty members outside of my department who wanted more courses in applied mathematics and statistics for their research. A few years later, this led to the formation of the Institute of Applied Mathematics and Statistics, in which I became a founding member.

I was lucky to be born a "pre-baby boomer" in 1940. In the 1960s, there were relatively few of us freshly minted PhDs looking for academic or research positions in Canada. By the 1970s, Canadian universities were being flooded with applications from highly qualified baby boomers born after 1945 and Viet Nam war avoiders from the US. In my case, after finishing a postdoc at Harvard in 1967, I had the luxury of three offers: a research scientist position with the Canadian Defence Research Board in Victoria, an assistant professorship in Mechanics at the prestigious Johns Hopkins University (JHU) in Baltimore (America's first research university founded in 1876), and an assistant professorship in Mathematics at the University of British Columbia (UBC) in Vancouver.

I ruled out the first option since I really wanted to teach as well as do research, and I turned down the second because I wanted to return to Canada. Among other issues, my wife Diana and I did not enjoy hearing daily news about the Viet Nam war. Thus, in August 1967, we (including Jasper, our large black poodle) made our way to Vancouver by car for me to join the UBC faculty.

It wasn't a smooth start. We stayed with my Uncle Peter (the father of my mathematician cousin Allan Trojan who was a professor at McGill) until we found our own apartment, which in hindsight was a mistake: Jasper picked up fleas in uncle's apartment! Then, shortly after moving into an apartment in the Kitsilano beach district, we had to leave because the landlord later decided he didn't want dogs. In October,

L. Mysak, *Adventures in Climate Science, Ocean Waves, and the Flute*, Springer Biographies, https://doi.org/10.1007/978-3-032-19848-8_10

the autumn rain started and continued for 19 depressing days. In addition, Diana was unhappy at having had to turn down an offer to do a PhD in English literature at JHU.

My first year of teaching was challenging. Every Monday, Wednesday and Friday at 8.30 am, I started tediously grinding through an advanced, very theoretical calculus class for honors mathematics and physics students. They were sleepy and looked bored; I often felt that they would rather be at the cafeteria eating UBC's yummy cinnamon buns and drinking hot coffee. And me too! My rowdy engineering students, whom I was trying to teach how to solve differential equations, tossed paper airplanes behind my back when I was at the black board. However, after I showed them real-life applications of these equations, they settled down.

My second year of teaching went much better. "Tell the engineers some jokes," my department head, Ralph James advised. "This will hold their attention." And he was right. Instead of teaching the advanced calculus course again, I introduced a new course on state-of-the-art advanced applied math methods, open to all graduate students in any science or engineering department. In addition to 10 registered students from chemistry, geophysics, mathematics, oceanography and physics, I was delighted that three mechanical engineering faculty members and my math department colleague George Bluman also attended my classes. They were all keen to learn about these methods for their research. This new course was a big hit across the university and continued to have healthy enrollments for many years.

It was also the year that I first supervised a graduate student, Mick Manton, who worked on oceanic internal waves [1]. During my 10 years at UBC, I supervised 9 PhD and 7 MSc students in applied mathematics and ocean dynamics. Of these, 8 became professors in Australia, Canada, South Africa and Spain.

Diana enrolled in the PhD program in English literature at UBC, which meant we both spent the whole day on campus. So, Jasper came with us to the university. However, he did not enjoy staying in my office all day. Jasper loved to explore the campus, even on rainy days. Unfortunately, on one such day he entered Professor Jame's large first-year calculus class and vigorously shook the water out of his stringy black hair all over the blackboard and James himself. Next day, not surprisingly, a short letter appeared in my mailbox. "In the future, please make arrangements to keep your big dog at home," Professor James wrote. This turned out to be impractical, so in the end, we asked my sister Helen in Toronto to take Jasper into her family. Happily, she agreed, and Jasper then had many happy years living with Helen and her children, Greg and Daphne.

In my third year (1969–70), George Bluman and I were both keen to discover what other departments in the university used in the way of mathematics to advance their disciplines. Through meetings with many individual faculty members, we soon learned more specifically what applied mathematics and statistics courses would benefit their research work. This inspired us and some statistics colleagues to develop a wide range of new courses in applied mathematics and statistics. This spade work also led to the formation of the Institute of Applied Mathematics and Statistics (IAMS) in the early 1970s, and George and I were founding members. The Institute's goal was to promote cross-disciplinary research in fields where mathematics and statistics played a fundamental role. At the time of writing this memoir, the Institute (now just IAM, as Statistics is now a separate department at UBC) is still very active.

In 1970, I was invited to become an associate member of the Institute of Oceanography, which was based in the Faculty of Graduate Studies. Since I had been doing research on various theoretical ocean wave problems for several years, I welcomed this appointment, which resulted in a collaboration with institute member Paul LeBlond, a physicist [2]. This collaboration continued when a few years later we were asked to write a comprehensive textbook on ocean waves (Chap. 13). Through my membership in the institute, I was also able to recruit new graduate students in physical oceanography.

The year 1970 was also notable for several other reasons. In May, when I was collecting my mail in the departmental office, Professor James called me over to the side. "The Dean of Science has just approved your promotion to associate professor," he whispered softly. "This means you can apply for sabbatical leave for the academic year 1971–72." I was delighted by this news. At that time, UBC had the policy that one could take a sabbatical after four years of teaching with 60% of your salary, tax free (thanks to the Federal Government). Thus, I soon made a formal application to the university to spend the academic year 1971–72 at the renowned Department of Applied Mathematics and Theoretical Physics (known as DAMTP) at the University of Cambridge, England.

I chose this department for a sabbatical visit because DAMTP was known to have many visitors each year who worked in physical oceanography and geophysical fluid dynamics (GFD). In addition, I knew a few DAMTP faculty who worked in GFD, including the widely esteemed Professors Longuet-Higgins and James Lighthill; the latter held a mathematics chair once occupied by Newton. In addition, I believed spending a year living in Cambridge would be a very broadening experience, as prior to this I had only made two short conference trips to Europe. But I was unaware that DAMTP is also where I would meet the famous cosmologist, Stephen Hawking.

In 1970, I welcomed Professor Ted Buchwald, from the University of New South Wales in Australia, as my first senior visitor to our department. We met briefly at Harvard when I was working on continental shelf waves [3] for my PhD thesis. As he and a student had recently written a pair of elegant papers on the generation and propagation of these waves [4, 5], Ted was looking forward to collaborating with me on some related ocean wave topic while visiting UBC. We began investigating the scattering of Rossby waves by ocean islands, a topic which I continued working on when I started my sabbatical at Cambridge in fall 1971. Unfortunately, we never completed this project due to Ted's heavy administrative load back in Australia and my new interest in wave propagation in random media. However, this did not discourage me from inviting other visitors to my department. During my two decades at UBC, I supervised about a dozen postdocs and hosted another dozen senior researchers, which reflected a growing interest in my research work on applied mathematics and physical oceanography. I am very proud of several papers I published with the latter group, which covered a multitude of topics ranging from the stability of coastal currents [6, 7], wave propagation in random media [8–10], the generation of Hawaiian Ridge currents [11], and trench wave generation and propagation [12].

The academic year 1970–71 was also a very challenging year for my wife Diana and myself. We were not getting on well, and I did not see a happy future ahead. In fact, two days before our wedding in 1965, Diana wanted to pull out. However, I convinced her to give marriage a try. But after five years of many ups and downs, I decided that we had to separate. In May 1971, I moved out of the house that we had been renting for the past three and half years. Fortunately, I was able to stay with Uncle Peter and his new wife Betty for the next few months while I prepared for my sabbatical visit to Cambridge.

After moving out of the house, I stored my personal belongings with Peter and Betty, and I gave most of the furniture and the car (a 1970 Volvo) to Diana. I agreed to give her half of my sabbatical salary during the next academic year and helped her find a new apartment.

I flew to London, England on August 1, 1971. It was a lonely day for me, but I did feel it was the right decision to have separated from Diana.

10.1 Epilogue

Diana went on to complete her PhD in English literature in the mid-1970s and then had a notable career as an English professor in southern British Columbia. We got officially divorced in 1974. I have had no contact with her since then.

Acknowledgements I thank my wife Janet Boeckh for her comments on the first draft of this memoir, George Bluman for his input, and Josef Schmidt for his helpful suggestions on a later version.

References

1. Manton MJ, Mysak LA (1971) Construction of internal wave solutions via a certain functional equation. J Math Anal Appl 35:237–248
2. Mysak LA, LeBlond PH (1972) The scattering of Rossby waves by a semi-infinite barrier. J Phys Oceanogr 2:108–114
3. Mysak LA (1967) On the theory of continental shelf waves. J Mar Res 25:205–227
4. Adams JK, Buchwald VT (1969) The generation of continental shelf waves. J Fluid Mech 35:815–826
5. Buchwald VT, Adams JK (1968) The propagation of continental shelf waves. Proc R Soc London A 305:235–250
6. Mysak LA, Schott F (1977) Evidence for baroclinic instability of the Norwegian Current. J Geophys Res 82:2087–2095
7. Niiler PP, Mysak LA (1971) Barotropic waves along an eastern continental shelf. Geophys Fluid Dyn 2:273–288
8. Howe MS, Mysak LA (1973) Scattering of Poincare waves by an irregular coastline. J Fluid Mech 57:111–128
9. Manton MJ, Mysak LA (1976) The stability of inviscid plane Couette flow in the presence of random fluctuations. J Engineering Math 10:231–241

10. Mysak LA, Howe MS (1976) A kinetic theory for internal waves in a randomly stratified fluid. Dyn Atmos Oceans 1:3–31
11. Mysak LA, Magaard L (1983) Rossby wave driven Eulerian mean flows along non-zonal barriers, with application to the Hawaiian Ridge. J Phys Oceanogr 23:1715–1725
12. Holyer JY, Mysak LA (1985) Trench wave generation by incident baroclinic Rossby waves. J Phys Oceanogr 15:593–603

Chapter 11
Making Waves in Cambridge, England

Abstract My first sabbatical was spent in the Department of Applied Mathematics and Theoretical Physics (DAMTP) at the University of Cambridge during 1971–72. I was on my own, having just separated from my first wife Diana. I was at a loss in my academic and social life. However, I soon met several new colleagues at the university and got involved in various musical, social and sporting activities. I discovered that there were a lot of young women in Cambridge, and by hanging around the University Centre I was able to find dates for concerts, plays, punting and dinners.

DAMTP had a lively seminar series in fluid mechanics which hosted speakers from around the world, many of whom I had read about in scientific journals. While having afternoon tea before one of these seminars, I was fortunate to meet applied mathematician Michael Howe from Imperial College, London, who shared a common interest in wave propagation in random media. Over the next few years, we published three papers on different types of ocean waves in the presence of random media or boundaries.

A turning point in my sabbatical occurred during a February camping holiday in Morocco when I met photographer Mary Burr, who lived in London. We were attracted to each other, and over the next six months we had many exciting adventures together, including a two-week road trip through Russia and Ukraine. In October 1972 Mary joined me in Vancouver, and after my divorce two years later, we married in 1974.

"This is Cambridge," the cultured woman's voice announced in the carriage as the train slowly rolled into the station. It was August 1971 and overcast and cool, but this did not dampen my spirits as I was about to begin one of the most fulfilling years of my life: new places, new friends and research colleagues, a return to my flute, new sports, a new woman in my life, and new tastes and smells. I couldn't wait to have my first plate of fish and chips and a pint of bitters in a Cambridge pub. But first I had to settle in my new lodgings, Flat 3, 1 Latham Road, on a quiet tree-lined residential street just south of the Department of Applied Mathematics and Theoretical Physics (DAMTP), where I would be spending the next academic year.

L. Mysak, *Adventures in Climate Science, Ocean Waves, and the Flute*, Springer Biographies, https://doi.org/10.1007/978-3-032-19848-8_11

11.1 The Flat

I was very lucky to find a flat thanks to the Cambridge Society for Visiting Scholars, which I had learned about from previous visitors to DAMTP. The flat was on the second floor of an elegant English manor home which had been divided into six self-contained units. From a bay window at the back, I had views of a multi-colored rose garden, and in the distance, the green playing fields of the prestigious independent Perse School Cambridge. I ate my meals in this peaceful nook and wondered what might be in store for me during this sabbatical year. One lazy Saturday afternoon, I took a stroll in the Perse School playing fields and was suddenly accosted by a well-dressed gentleman and his Irish Setter. "Are you lost?" he politely asked. I quickly apologized, realizing that I was not welcome to walk on the school's private grounds, and hastily returned to my flat.

For a one-bedroom unit the flat was huge. The living room area had two sofas opposite a large coffee table, a dining room suite and sideboard, an out-of-tune piano, three easy chairs, and a spare bed. The kitchen was small, however, with a tiny fridge, limited counter and cupboard space, and a two-burner gas stove. But it was sufficient for me to prepare simple meals. The flat was ideal for entertaining, which I did a lot of during the year.

In 30 min, I could walk to DAMTP along the winding River Cam. Shortly after my arrival I bought a second-hand bicycle, which shortened my commute to 10 min. Surprisingly, I was able to cycle most days to the office as there was relatively little rain and no snow during the year I was there.

11.2 DAMTP

Fluid dynamics specialist George Batchelor [1] founded DAMTP in 1959 when he was only 39, and he continued as Head of DAMTP until 1983, when it was taken over by magneto geodynamics expert Keith Moffat. In its early years, the department was internationally renowned in two main areas: fluid and solid mechanics, and cosmology and general relativity. When I was a "Senior Visitor" there in 1971–72, DAMTP had many faculty and researchers in the rapidly developing field of geophysical fluid dynamics [2–4], which is one of the reasons why I chose Cambridge for my first sabbatical. With 30 faculty members and 40 PhD students combined with the many tourist attractions of Cambridge, it was not surprising that DAMTP had many academic visitors. At one point, I counted 25 visiting postdocs and professors, like me. It was an extremely stimulating scholarly environment, and I never wanted to miss the morning coffee and afternoon tea sessions in the large lounge which once housed the printing machines of the Cambridge University Press. Among the distinguished personalities I met that year were Stephen Hawking, Sir Harold Jeffreys, and Sir James Lighthill.

In the fall term I attended a set of lectures on waves by Lighthill [3], who was the Lucasian Professor of Mathematics, a position once held by Newton in the late 1600s. He began with Newton's theory of sound waves in an atmosphere of constant density and showed that the computed sound speed did not agree with the observed speed. "The reason for this," Lighthill explained, "was that my "predecessor" Sir Isaac Newton did not realize that the air density varies with the air temperature." So, not everyone gets it right the first time!

George Batchelor was very welcoming to visitors, but he had a dry sense of humor. He made it clear that because of limited secretarial staff, visitors had to do their own typing of manuscripts and letters on an old typewriter the department provided. He also pointed out that paper was expensive in England. When I asked him where I might find some paper for my mathematical calculations, he responded, "How many pages do you need per day?" When I replied, "About a dozen," he said I could raid the basket of discarded copies by the Xerox machine!

In September Tony Chen, on sabbatical from Rutgers State University, New Jersey, joined me as an office mate for the year. He did both experimental and theoretical work on mixing in stratified fluids and planned to work with Stuart Turner [4]. We became close friends over the year and shared many lunches in the University Centre and the pubs of Cambridge. We especially liked having oily fish and chips at The Eagle, where in the early 1950s Watson and Crick had discussed ideas which led to the discovery of the double helix, the twisted-ladder structure of deoxyribose nucleic acid (DNA), described in a one-page note in the 25 April 1953 issue of Nature.

11.3 Research and Scientific Presentations

I came to Cambridge with the intention of continuing my research on stochastic differential equations and ocean wave propagation in random media, which I had started at UBC. During winter I was fortunate to have met Michael Howe from Imperial College London, who was visiting DAMTP to consult Lighthill about acoustic wave scattering from an irregular boundary. I quickly realized that the mathematical approach he used could be applied to the scattering of ocean tidal waves by an irregular coastline [5]. Howe also became interested in my new work on internal waves in a randomly stratified fluid [6, 7], which led to an elegant joint paper on a kinetic theory approach to describe energy transport in the ocean's interior [8]. As I had hoped, coming to Cambridge widened my scientific network and introduced me to new areas of research.

The southern part of England has many universities and research institutes where I gave seminars on my earlier work on Rossby waves [9], shelf waves [10] as well as on my recent work on internal waves. In addition to speaking in DAMTP twice, I gave seminars at Imperial College London, Oxford University, University College London, the universities in Essex, Reading, Surrey and Warwick, and the oceanography institutes in Wormley and Liverpool. Since I had a car in England (a new Volvo), I drove to many of these locations and hence enjoyed seeing the ever-varying English countryside, which seemed to have green grass in all seasons. These visits certainly gave my research work considerable exposure and helped establish my reputation as an upcoming theoretical oceanographer who specialized in ocean waves.

I attended two stimulating international conferences in Europe during my sabbatical. These visits also led to new contacts and future collaborators in my field. And I tasted new foods and heard new languages. In March 1972 I gave an invited paper on internal waves in a randomly stratified fluid [6] at the Fourth Colloquium on Ocean Hydrodynamics in Liège, Belgium, and in August I gave the same paper at the 13th IUTAM Congress in Moscow. At both conferences delegates were puzzled by the theoretical result that random fluctuations in the stability frequency cause a growth (decay) in the amplitude of a coherent wave whose phase propagates upward (downward). A year later, my postdoc Charles Tang fortunately found a mathematical error in the calculations in [6]—an incorrect Green's function was used. When the correct Green's function was implemented [7], the abovementioned result was discovered to be fallacious and it was shown that in the direction of energy propagation, a coherent wave is always attenuated because of the energy transfer to the incoherent wave field.

11.4 Music: Revisiting the Flute and Learning the Guitar

Cambridge is a musical city which also has many cultural societies. While the University of Cambridge is renowned for its science, it has an excellent music school. One of its young lecturers conducted the amateur town orchestra known as the "Cambridge Philharmonic Society". I had not played the flute for several years, and it took me a month to get my flabby embouchure back into shape after I arrived in Cambridge. In October, I went to the orchestra's first rehearsal at the suggestion of a neighbor and had a short audition with the conductor. I was delighted that he welcomed me into the ensemble, as did the longstanding flute player who during the day ran a high-end haberdashery in central Cambridge. It was a pleasure to play in an orchestra again. As well as participating in three orchestra concerts, I played in the orchestra pit for the Cambridge Opera Society's performances of Offenbach's *Tales of Hoffmann* and in an ensemble accompanying a presentation of Bach's *St. Matthew Passion.*

In October I also started classical guitar lessons on a golden-colored Spanish-made instrument which I rented for two quid a week from a music shop in Cambridge. I was inspired to take these lessons after hearing a touching and evocative performance by Julian Bream, who played both the guitar and lute at his concert. Making gentle sounds on the guitar was very soothing for me during the long evenings. In July I ended up buying the guitar by taking advantage of the hire-purchase scheme that the music shop offered. I continued to play the guitar when I returned to Vancouver in September 1972, but I soon had to quit because the long fingernails needed on the right hand for plucking the guitar strings made it very difficult to play the flute. And I really wanted to continue with my flute playing when I returned to Vancouver. I eventually gave the guitar to my daughter Claire, who as a teenager was a gifted singer and guitar player.

11.5 Sporting Activities

Punting on the River Cam is an excellent way to see the "'backs" of the historic colleges of the University of Cambridge. The "Bridge of Sighs", built in 1831, is an iconic feature connecting the two parts of St. John's College (Photo 11.1). It is best seen from a punt, which I learned to use shortly after arriving in Cambridge.

However, punting is far from easy and requires an excellent balance when standing on a small platform at the rear of the boat and pushing a long pole into the soft riverbed (Photo 11.2). I was afraid of falling into the river and hence always wore minimalist clothing when punting (Photo 11.2).

Before coming to Cambridge, I was lucky to have met Lesley when she was an exchange teacher in Vancouver. She not only showed me many of the tourist highlights of Cambridge after my arrival in the city, but also taught me how to play squash, which is a very popular racket game in England. I loved the speed and skill required for this game and continued playing until my mid-50s. By this time,

Photo 11.1 Bridge of Sighs across the River Cam in Cambridge, England

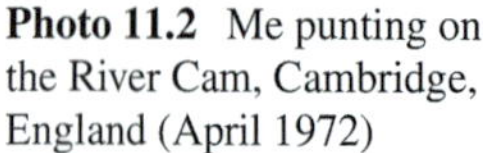

Photo 11.2 Me punting on the River Cam, Cambridge, England (April 1972)

however, my feet could not move as quickly as my mind, and I was prone to getting many twisted ankles. So, I switched to tennis, which I enjoyed playing until my early 80s.

Being a sailing nation, England offered many opportunities for sailboat rentals. Early in the fall, my UBC colleague Jim Zidek (who was on sabbatical in London), his wife Lynne and I charted a 25-foot yacht for a weekend sail on the Blackwater River on the east coast of England. Rather than docking in a harbor overnight, it is customary to drop anchor in the middle of the river and let the boat's keel sink deep into the soft mud as the tide went out. Since the tides there are semi-diurnal, the boat becomes afloat again when the tide returns the next morning. Buoyancy is a powerful force!

In the late spring, we experienced the mud of another English river that flows into the Solent, on the south coast of England. In this case, as we sailed up the Beaulieu River at high tide, we were supposed to follow a course between two parallel lines of poles, which marked a deep-water passage. Unfortunately, we lost our way and were soon sailing ever so slowly through the mud as the tide went out (Photo 11.3). We soon came to a dead halt. However, in the middle of the night when the tide was coming in, we floated again and managed to motor out to deeper water. I learned that sailing in England is so different from the West Coast of Canada where one is more concerned about crashing into large rocks and boulders rather than getting stuck in the mud.

Photo 11.3 "We should have sailed there," I was telling my friend Jim Zidek after we got stuck in the mud of the Beaulieu River on the south coast of England (May 1972)

11.6 My Social Life in Cambridge

Being a bachelor again was a lonely experience, especially on a weekend. To help overcome this, I went to many concerts and plays, in some cases with women friends I met at the University Centre. Rowena, a Cambridge PhD student in English whom I met a year earlier in Birmingham when I visited UBC friend David Short there, was a vivacious redhead who introduced me to several of her female friends. I had a crush on her, but she was still "attached" to David. On the positive side, she did teach me how to cook delicious pork and chicken casseroles with wine, a skill I used later when entertaining couples and women friends in my flat. When she broke up with David later in the fall, I had hopes that she might be the new woman in my life. But she quickly gave me the brushoff, maybe because she knew I was unhappily married to another English scholar in Vancouver.

During the first six months in Cambridge, I had a very active social life. I dated over half a dozen women, the last one being Judy, a teacher who lived upstairs in another flat at 1 Latham Road. The relationship got quite serious. In addition to going to many concerts and plays, I met her parents over a roast beef dinner at a fancy Cambridge hotel and accompanied her to the annual school ball. But the relationship ended when I went on a camping holiday in Morocco at end of February and met photographer Mary Burr, who happened to take the same Mini-trek expedition in the southwest corner of this country (Chap. 12). Mary lived in London, a one-hour train ride from Cambridge.

During the second half of the sabbatical, Mary came to Cambridge on most weekends. She enjoyed theatre and music and was a talented pianist; we played piano and flute duets in my flat. Mary also had a great sense of adventure, having lived in Australia and Kenya for several years. Among other activities, she joined the Zideks and me on a weeklong sail in the Solent, where we ran aground in the mud (Photo 11.3). That sailing trip also had a very dramatic beginning. We crossed the Solent from Chichester Harbor to Isle of Wight in a Beaufort scale 9 gale. We were lucky to have made it across without a broken mast, which happened to many others on that crossing. As we were hanging on for dear life, I could see the fright in Mary' s eyes. "Have you had much experience sailing in these conditions? she timidly asked. "Not much," I replied, as I gritted my teeth in the hope of a safe crossing.

11.7 Personal Travels

During my time in Cambridge, I was delighted to host my uncle Peter Trojan and his wife Betty (in August 1971) and my mom and dad (in July and August 1972). Peter and Betty came to England from Vancouver shortly after my arrival in Cambridge to attend the wedding of my mathematician cousin Allan Trojan (Photo 2.11, Chap. 2). Allan married Diana Hobbins, who worked in the library at McGill University where Allan was a young professor. The Hobbins' family lived in Bolton (near Manchester), where Diana had grown up. We travelled there by train. It was a very formal wedding, with all the local gentleman in morning suits; in contrast, I wore a bright green and yellow striped sports jacket which clearly showed that I was an out-of-towner. Diana's brother John, who is now retired from McGill and is one of my current bridge buddies, has never let me forget my inappropriate attire at his sister's wedding. But we are now good friends, and he calls me "Cousin Lawrence".

After the wedding, Peter, Betty and I flew to Amsterdam where I picked up a new Volvo which I had ordered in Vancouver, on the European Delivery Plan. Because I was going to be driving this car in Europe and England for over a year and then bring it back to Vancouver as a used car, I paid about 30% less than the price of a new Volvo at home. After touring Amsterdam, I took Peter and Betty on a road trip to Geneva, Paris, Brussels and Brugge. I used the Harvard student handbook "*Let's Go Europe*" to find economical hotels and sites of interest in each of the cities. Since most of our travels were in French speaking areas, we had some challenges with the local French dialects. But we survived and made our way back to Cambridge on the Oostende (Belgium) to Harwich (England) overnight ferry. When Uncle Peter saw the view of the green fields and tall Lombardypoplar trees from the bay window of my flat, he said, "Lawrence, I think you will have a peaceful and wonderful year in Cambridge." And he was right.

Mom and Dad retired as teachers on June 30, 1972, and then flew to London from Edmonton in mid-July for their first-ever trip to England and continental Europe. After seeing the highlights of London (including feeding the pigeons in Trafalgar Square) and meeting my new friend Mary Burr (Photo 11.4), we drove to Bolton and Newcastle to visit in-laws and spent a couple of nights in historic Edinburgh. We then took the ferry from Harwich to Oostende for a 10-day tour of Western Europe. They never forgot this trip, and we often talked about the tasty meals we had in Brussels, Geneva and Paris. Mom gave me her European diary before she died in 1978, and I loved the following entry she made on 5 August 1972: "La Rose Restaurant on Grande Place, Brussels. The supper was absolutely delicious. Steak was so tender. Best meal so far. Very expensive." Yes, we agreed, top food places are not cheap.

Mary flew into Brussels on our final day there, and after seeing Mom and Dad off at the train station (they would go to London and then fly back to Edmonton), Mary and I started off on a most challenging and exciting adventure: a two-week road trip through Russia (entered from Helsinki) and Ukraine (all the way to the Romanian border), and then back through Ukraine and Russia, ending up in Moscow. I often had to struggle with my elementary Russian and Ukrainian, but it was enough to get us through most awkward situations. We were always under surveillance in Russia and Ukraine, and finding our way with very little road signage was not easy. Nevertheless, we made it back to Moscow where Mary flew home to London (with great relief) and I began my participation in the IUTAM Congress (see above section, Research and Scientific Presentations).

Some details of our 1972 road trip though Russia and Ukraine are described in Chap. 32, along with short accounts of our 1993 and 2002 trips to Ukraine. One day I will write a longer memoir about my travels and experiences in Ukraine.

Photo 11.4 Mary Burr at a wedding reception in London when I was touring Europe with Mom and Dad (July 1972)

11.8 Epilogue

I returned to DAMTP in May 1977 to give an invited seminar on "Coastal Trapped Waves: A Review of Results Old and New", which was later published in the Reviews of Geophysics and Space Physics [11]. The audience in the seminar room was as lively as I remembered five years ago; I was continually peppered with questions by two brilliant young fluid mechanicians, Michael McIntyre and Hubert Huppert. They are both now FRS. Mary and I, together with our two-year old son Paul, were invited to an informal dinner party hosted by the Longuet-Higgins family. It was great to see so many old colleagues again.

Today DAMTP is no longer situated on Silver Street near the centre of town, as was the case in 1971–72. It moved in 2002 to the countryside west of Cambridge and is now part of the Centre for Mathematical Sciences, which also houses the Department of Pure Mathematics and Mathematical Statistics, and the Isaac Newton Institute. DAMTP has also grown enormously in size, to 140 professorial and research staff and 150 postgraduate students. It still hosts over 30 visitors every year. The new areas of research carried out in DAMTP include environmental fluid mechanics, mathematical biology, and quantum information theory. It is still well known internationally, however, for its work in applied computational analysis, astrophysics and geophysics, fluid and solid mechanics, high energy physics, and cosmology and relativity.

Looking back on my year at Cambridge more than 50 years later, clearly, I blossomed there internationally as a young Canadian scientist working in ocean waves. Returning to my silver flute in Cambridge was the start of over 50 years of enjoyment playing orchestral and chamber music. And I will always be grateful that I met Mary Burr in Morocco in February 1972 and that she joined me in Vancouver in October. We were together for 40 years and had two wonderful children. The next chapter gives the details about our meeting and early relationship.

Acknowledgements I thank Janet Boeckh for her comments on the first draft of this memoir, Claire Mysak for inserting the photographs, and Josef Schmidt for his helpful critique.

References

1. Batchelor GK (1967) An introduction to fluid dynamics. Cambridge University Press, London
2. Gill AE (1982) Atmosphere-ocean dynamics. Academic Press Inc., New York
3. Lighthill J (1978) Waves in fluids. Cambridge University Press, London
4. Turner JS (1973) Buoyancy effects in fluids. Cambridge University Press, London
5. Howe MS, Mysak LA (1973) Scattering of Poincare waves by an irregular coastline. J Fluid Mech 57:111–128
6. McGorman RE, Mysak LA (1973) Internal waves in a randomly stratified fluid. Geophys Fluid Dyn 4:243–266
7. Tang CL, Mysak LA (1976) A note on "Internal waves in a randomly stratified fluid." J Phys Oceanogr 6:243–246
8. Mysak LA, Howe MS (1976) A kinetic theory for internal waves in a randomly stratified fluid. Dyn Atmos Oceans 1:3–31
9. Mysak LA, LeBlond PH (1972) The scattering of Rossby waves by a semi-infinite barrier. J Phys Oceanogr 2:108–114
10. Niiler PP, Mysak LA (1971) Barotropic waves along an eastern continental shelf. Geophys Fluid Dyn 2:273–288
11. Mysak LA (1980) Recent advances in shelf wave dynamics. Rev Geophys Space Phys 18:211–241

Chapter 12
Wooing Mary and Winning Over Mom

Abstract In the summer of 1972, I took Mom and Dad on a road trip in England, Scotland and western Europe. Before leaving, I introduced them to Mary Burr whom I had met on a camping holiday in Morocco and had been seeing for the past six months. This was an awkward situation for my parents since I was still married although separated from my first wife. My mother was hoping that I would later get divorced and meet another Ukrainian woman. I convinced her that Mary was the woman for me and told her that Mary would be joining me in Vancouver in the fall.

Mom and Dad exit Customs at Heathrow with smiles on their faces. They are excited about their first trip to Europe in July 1972. I'm taking them to Cambridge, where I'm spending a sabbatical in the university's Department of Applied Mathematics and Theoretical Physics. "Before driving to Cambridge," I said, "I want you to meet Mary."

Mom frowns. "Who is Mary?" she asks.

"A woman I met in Morocco last February. She lives in London, and we've been seeing each other since March. She's a photographer."

I came to Cambridge in August 1971 after going through a painful separation from Diana, who is Ukrainian Canadian like me. We started a relationship in 1964 while I was teaching summer school at the University of Alberta in Edmonton. In the 1960s, Ukrainian boys generally married Ukrainian girls. Diana and I married in a huge, multi-domed Ukrainian Catholic Cathedral in 1965. I was still in graduate school at Harvard, and two years later, we moved to Vancouver, where I began teaching at UBC (Chap. 10).

I walked out of the marriage in spring 1971 because during the previous couple of years we were not getting on very well; I did not see a happy future. I developed stomach and back pains from the stress of trying to make our marriage work. When my sabbatical application was approved by UBC, I decided to go to Cambridge alone.

When I was in Cambridge, I had to learn the dating game all over again. And it was not easy to tell an attractive woman that I was legally married but recently separated.

L. Mysak, *Adventures in Climate Science, Ocean Waves, and the Flute*, Springer Biographies, https://doi.org/10.1007/978-3-032-19848-8_12

That fall I had a couple of brief affairs in Cambridge. But none lasted. Then in January, a newspaper ad caught my attention. "Mini-trek Expeditions to North Africa" the headline stated and below was pictured a land rover in an oasis with people unloading camping gear from the roof rack. Looks like fun I said to myself.

At the end of February, I was the one unloading camping gear in Agadir, the starting point of a two-week trip to the southwest part of Morocco.

Richard, our expedition leader, was on the roof of the land rover when he lost his balance and fell on top of Mary. He flattened her to the ground. But she quickly got up, dusted herself off and said, to Richard's great relief, "I'm okay." Wow, Mary's one tough cookie I thought.

Mary had a photographer's eye. In the evening after dinner, a group of us went to the harbor to see the sunset. "Look at the silvery moon shining over the breaking waves crashing onto the silvery sandy beach," she remarked. "This could be a great photograph."

The rest of us looked at each other and wondered, who is this woman?

Over the next six days, I fended off two fellow campers who also had their eyes on Mary. However, I was nervous when it came time to tell her of my marital situation. She did not know that I was already married but recently separated. When she heard this Mary asked two incisive questions. "Who left who, and do you have any children?".

"I left my wife last summer to go to Cambridge," I replied, and "no, I don't have any children."

That's all she needed to hear. I got light-headed kissing Mary in the high snow-peaked Atlas Mountains. Our romance blossomed.

Mary liked adventure, having lived on three continents. This was a *good omen* for our relationship as I also loved to travel. We met on weekends, either in London or Cambridge, and in June we had an exciting sail in the stormy Solent (Chap. 11). In August 1972, at the height of the Cold War, we had the adventure of our lives traveling by car and camping in Russia and Ukraine (Chap. 32).

But Mary was not Ukrainian. Her mother was English and her father a Scot.

When Mom and Dad met Mary in London that summer, Dad liked her energy and curiosity. But Mom didn't think our relationship would last because of our very different backgrounds. She kept saying to me privately, "I will find you a nice Ukrainian girl back in Edmonton."

"No, no, no, Mary is the woman for me," I replied very firmly.

Mary came to live with me in Vancouver in late October, six weeks after I returned from Cambridge in September 1972, and after 18 months of tough negotiations I got divorced. We married in August 1974, and by this time Mom had grown to like Mary. They got on well, and Mom taught her a few Ukrainian dishes. Mary and I were fortunate to have two healthy children, Paul and Claire, born in 1975 and 1978. Sadly, Mom died of cancer in 1978, and she only got to know Paul as a toddler. She never met Claire, who was born four days before her death. However, as Claire's middle name is Anastasia, which was my mother's first name, she is remembered to this day.

12.1 Epilogue

This chapter is a revised version of an assignment which I was given in the Thomas More Institute (TMI) Memoir Writing Course that I took in winter 2018. I am very grateful to Pauline Beauchamp and Karen Nesbitt, the course instructors, for their feedback and suggestions for the improvement of this short memoir.

Chapter 13
Writing *Waves in the Ocean*, a Serendipitous Project

Abstract In the fall of 1973, a managing editor from Elsevier Scientific Publishing Company came to UBC's Institute of Oceanography looking for Paul LeBlond to write a book on ocean waves. As Paul was away on sabbatical, I was then approached to take on this task. In the end, Paul and I agreed to do this project jointly. It took us three years (1974–77) to complete the 600-page book, which appeared in the spring of 1978. Despite its high price tag, the book was a great success with both researchers and students because of its extensive coverage of all classes of ocean waves and their interactions with each other and their environment.

13.1 How It Started

In October 1973, I got a call from Rolf Käse, a visiting postdoc at UBC's Institute of Oceanography. "Lawrence," he said, "I have a gentleman in my office who is looking for Paul LeBlond, but I thought you might be interested in what he has to say." Rolf was from Germany and was temporarily using Paul's office while he was on sabbatical in Moscow during 1973–74.

The gentleman, whose name I believe was Herman Frank, was a managing editor from Elsevier Scientific Publishing Company in The Netherlands. He had come to UBC to entice LeBlond to write a textbook on the theory of ocean waves, a topic that Paul had taught and researched. Because Frank had already been turned down for this project by more senior oceanographers, such as Walter Munk NAS and Michael Longuet-Higgins FRS, he was now making the rounds with younger physical oceanographers in the hope of getting a commitment to write such a book for the Elsevier Oceanography Series. Rolf, who had a quick mind, thought I might be interested in doing this and promptly brought Frank to my office in the Mathematics Department.

My first reaction to Frank's inquiry was that Paul should first be given the chance to write it on his own. When I wrote to him that fall, he replied, "No, it'll be more fun to write a book together." I was delighted with his suggestion. Later Frank sent a contract for us to sign, which we did together in Melbourne in January 1974 when we

L. Mysak, *Adventures in Climate Science, Ocean Waves, and the Flute*, Springer Biographies, https://doi.org/10.1007/978-3-032-19848-8_13

were attending an IAPSO conference at the University of Melbourne (see Chap. 22). The contract called for a 300-page book which would cover the dynamics of all types of ocean waves (surface and internal, tidal, planetary, coastally trapped and equatorially trapped) and their interactions with each other and with the medium through which they travel. We agreed to deliver a manuscript to Elsevier by the spring of 1977.

13.2 Organizing the Book

To help guide our writing, we examined two recent 300-page books [1, 2] on the dynamics of surface and internal gravity ocean waves in a non-rotating fluid. However, these books focused on small-scale mixing and turbulence in the ocean, which would be of secondary interest in our book. Paul and I were expected to describe, in addition to surface and internal waves, many other types of waves in which rotational and boundary effects are important. Shortly after we began writing our book, Elsevier published *Waves in the Atmosphere* [3], a catchy title for a book on atmospheric gravity and ultrasound waves. We thus used a similar title for our proposed book, *Waves in the Ocean* [4]. In contrast to the book by Earl Gossard and William Hooke, however, we planned to include exercises at the end of each section so that our book would be useful as a textbook for both undergraduate and graduate students.

We prepared an outline for the book in April 1974 and started writing in the summer of 1974, after Paul returned to Vancouver. We decided to develop the theory of various types of ocean waves in domains of increasing geometrical complexity, starting with plane waves in an unbounded, rotating stratified fluid. We would then introduce the effects of boundaries and thus discuss the panoply of possible modes of oscillation of the real ocean. Next, we would describe the nonlinear interactions between waves of each class and then investigate the stability of mean flows. Finally, we would conclude with the generation and dissipation of the different wave classes. Because the topic of ocean waves was rapidly evolving, we planned to include as many recent theoretical and observational studies as possible.

13.3 Dividing up the Work

How did we decide which parts each of us would write? This was not straightforward. For the introductory chapter, on the governing Navier Stokes equations of fluid dynamics and the general properties of plane wave theory, we leapfrogged through its sections. This meant that over a one-week period, Paul and I would simultaneously do a draft of Sects. 13.1 and 13.2 respectively; then on the following Monday we would meet to read the other's draft and make revisions. For the next week, we would write Sects. 13.3 and 13.4 but alternate the order of authorship; I would write

Sect. 13.3 and Paul Sect. 13.4. Then we would meet on the following Monday to review these two drafts and make revisions. By working this way, we developed a common notation for the mathematical symbols and kept a similar level of scientific sophistication for each section. The revised sections were promptly typed up by various secretarial staff at UBC.

For the second and third chapters, on waves shorter than the ocean depth and waves longer than the depth respectively, we each wrote all the sections of one chapter (Paul did Chap. 2 and I did Chap. 3). Then, after writing each section of our respective chapters, we would meet on a Monday to exchange drafts and do revisions. For Chaps. 4 and 5, on propagation in wave guides and statistical/probabilistic methods respectively, we again used the leapfrogging method of writing as in Chap. 1. Paul wrote most of Chap. 6, on wave-wave interactions and nonlinear solitary waves, and I wrote most of Chap. 7, on critical layer absorption (energy loss to mean flows when the wave phase speed matches the mean flow speed) and the stability of mean flows. The final chapter, on wave generation and dissipation, we wrote as a joint effort, as in Chaps. 1, 4 and 5.

Paul had a brilliant mind and wrote quickly the weekly assigned section, even though it was in his second language, English, which he learned at McGill and UBC. (Paul was a native French speaker from Quebec, but later he also developed a working knowledge of German, Russian and Spanish.) For me it was more of a struggle to write the weekly text, and I was often up late on Sunday evening finishing the section I had agreed to write. But I always had something for Paul to read on the following Monday, even if it was not a very polished piece.

Paul and I were especially grateful to three people who carefully read and worked through the mathematics of several chapters of the book: Professor Len Todd, Dr. Richard Thomson, and graduate student Daniel Wright.

Len Todd, an applied mathematician from Laurentian University, came to UBC in the fall of 1975 to spend a sabbatical in my department. Rather than start a new research project with me, he agreed to read the first four chapters of the book, which were now available for review. He carefully dissected every paragraph of text and line of mathematics from the point of view of a professor teaching an introductory course on ocean waves.

Richard (Rick) Thomson, a scientist from the Institute of Ocean Sciences in Sidney, BC and one of Paul's former PhD students, read those draft chapters of the book that were of particular interest to him as an oceanographic researcher wanting to learn about the latest advances in ocean waves. We took great delight in his enthusiastic and humorous approach to this task.

Daniel (Dan) Wright, a 1975 graduate of Laurentian University, had won a prestigious NRC doctoral fellowship to study the application of mathematics to fluid mechanics at any Canadian university. I was very lucky that he came to UBC in September 1975 and wanted me as his PhD supervisor. By now Paul and I had finished a draft of about two-thirds of the book, and we were looking for an outstanding applied mathematics or physics graduate student to carefully read what we had written and to continue working with us as an editor until the manuscript would be sent to the publisher. In addition, we wanted a student to work through all the exercises at the end of each section. Dan fit the bill perfectly. As he was married with a young daughter, he appreciated the extra stipend as a research assistant for doing this work. In the end, Dan worked through two versions of the complete manuscript and, in so doing, helped to eliminate many errors and obscurities.

While working through the sections on the stability of mean flows, Dan said to me, "Hey, I really like this stuff. I want to study it more."

"Great," I replied, "maybe we can think of a thesis problem on this topic." And indeed, we did. Dan developed a stability theory for large-scale oceanic flows which he later applied to the Antarctic Circumpolar Current.

By January 1977, we had a well-edited version of the whole book, thanks to Len, Rick, and Dan. We then sent single chapters to a dozen senior colleagues in Canada and abroad for their comments. We greatly appreciated their help; their names are listed in the book's preface. On top of this, while on a seven-month (February to August 1977) sabbatical visit to the National Center for Atmospheric Research in Boulder, Colorado, my office mate Peter Gent asked to look at our sections on surface waves and their generation. He then told me of the latest research on the flow separation of surface waves, which we were able to include in the final manuscript. While I was away from UBC, Paul checked over the references, especially those written in Russian, and assembled the pictures and figures for the book, many of which were drawn by his talented cousin Sylvie Lacerte. The final typed manuscript, with double line spacing, was prepared by Maryse Ellis; it was over 1200 pages and included about 1000 references. We sent the final package to Elsevier Publishers in April 1977, precisely three years after we wrote the book's outline.

13.4 Publishing the Book

In early August 1977, Paul received the proofs of our book in Vancouver. While we were both attending a fluid dynamics conference in Seattle at the end of August, we began the lengthy task of checking them. It took us two months to do this. We were now anxious to see the final product, which came in the mail in spring 1978 (Fig. 13.1).

We were very pleased with the quality of the book—the text was easy to read, the equations looked quite elegant, and the figures were well produced. However, when the advertisement for the book came out (Fig. 13.2), we were shocked at the price—US $98.50. We attributed this to the size of the book (602 printed pages) and its many equations, which had to be typeset by hand.

The first review of the book appeared in August 1978, in the British journal *Nature*, and it was very positive. David Webb, a physical oceanographer at the UK Institute of Oceanographic Sciences, wrote, "The authors are to be commended on producing an excellent review of waves in the ocean, but they must be held partly responsible for the price, which puts the book beyond the reach of most readers." We had to agree with his latter statement, which limited the sales of the first edition to just over 1000 copies. A paperback version with corrections appeared in 1980 and cost around US $

Fig. 13.1 Book cover of *Waves in the Ocean*, published by Elsevier in 1978

50; I think another 1000 copies of the book were quickly sold. We did not get rich on the book's royalties, but we were delighted that many colleagues and students found the book timely and useful in their research and course work. At an International Union of Geodesy and Geophysics (IUGG) General Assembly in Canberra, Australia in 1979, we received many compliments from the delegates there. In 1981, Paul and I were awarded the CMOS (Canadian Meteorological and Oceanographic Society) President's Prize for the book. Also, in 1981 *Waves in the Ocean* was translated into Russian and published as a two-volume set by "Мир" in Moscow. Graduate students from China and South America have told me that the book has been translated into Chinese and Spanish, but I have never seen these copies.

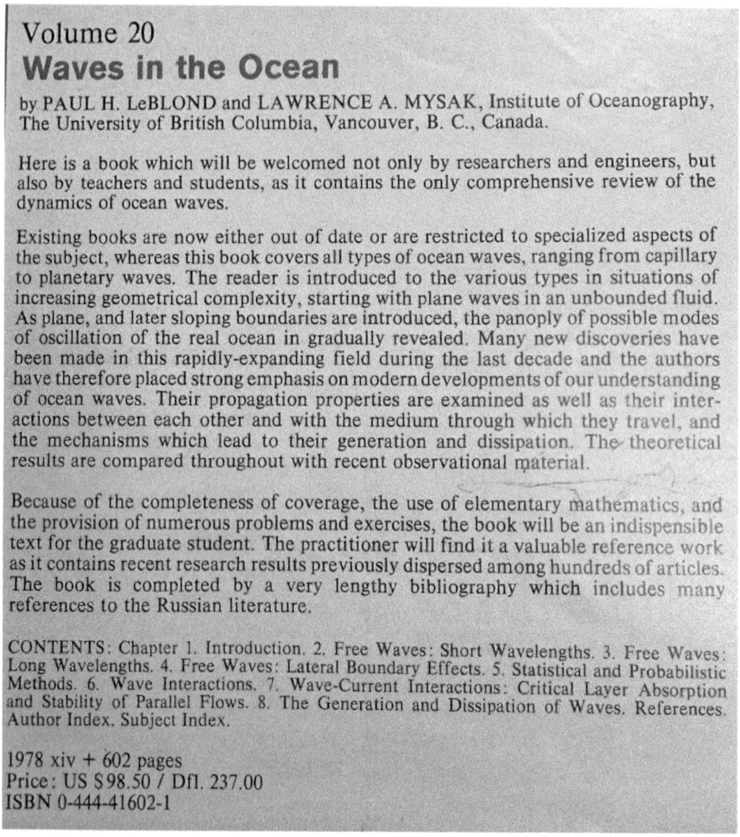

Volume 20

Waves in the Ocean

by PAUL H. LeBLOND and LAWRENCE A. MYSAK, Institute of Oceanography, The University of British Columbia, Vancouver, B. C., Canada.

Here is a book which will be welcomed not only by researchers and engineers, but also by teachers and students, as it contains the only comprehensive review of the dynamics of ocean waves.

Existing books are now either out of date or are restricted to specialized aspects of the subject, whereas this book covers all types of ocean waves, ranging from capillary to planetary waves. The reader is introduced to the various types in situations of increasing geometrical complexity, starting with plane waves in an unbounded fluid. As plane, and later sloping boundaries are introduced, the panoply of possible modes of oscillation of the real ocean in gradually revealed. Many new discoveries have been made in this rapidly-expanding field during the last decade and the authors have therefore placed strong emphasis on modern developments of our understanding of ocean waves. Their propagation properties are examined as well as their interactions between each other and with the medium through which they travel, and the mechanisms which lead to their generation and dissipation. The theoretical results are compared throughout with recent observational material.

Because of the completeness of coverage, the use of elementary mathematics, and the provision of numerous problems and exercises, the book will be an indispensible text for the graduate student. The practitioner will find it a valuable reference work as it contains recent research results previously dispersed among hundreds of articles. The book is completed by a very lengthy bibliography which includes many references to the Russian literature.

CONTENTS: Chapter 1. Introduction. 2. Free Waves: Short Wavelengths. 3. Free Waves: Long Wavelengths. 4. Free Waves: Lateral Boundary Effects. 5. Statistical and Probabilistic Methods. 6. Wave Interactions. 7. Wave-Current Interactions: Critical Layer Absorption and Stability of Parallel Flows. 8. The Generation and Dissipation of Waves. References. Author Index. Subject Index.

1978 xiv + 602 pages
Price: US $98.50 / Dfl. 237.00
ISBN 0-444-41602-1

Fig. 13.2 Publisher's advertisement for *Waves in the Ocean*. Permission granted by Elsevier Scientific Publishers

13.5 Friendships

During the writing of the book, Paul and I developed a close friendship, and we rarely had any serious disagreements over how or what to write in the book. Paul always wanted to be precise in the discussion of the physics of wave motions, and I always tried to be straightforward about the mathematical derivations. Thus, we built on each other's strengths.

While writing the book, Paul became the godfather of my son, Paul Alexander Mysak, born in June 1975 (Photo 13.1). Also, our families got to know each other well during those years, and we often spent holiday time together (Photo 13.2).

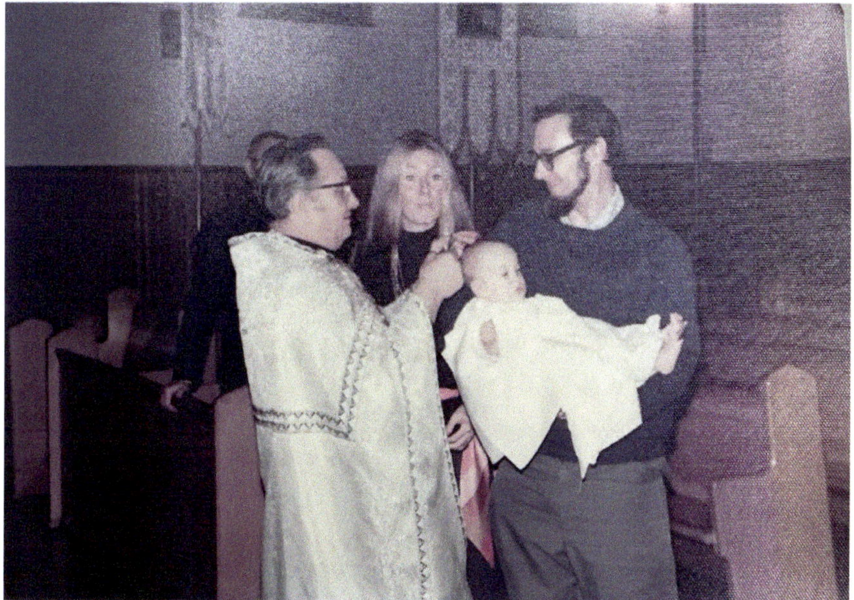

Photo 13.1 Paul LeBlond (right) holding my son Paul Mysak at his Christening in the Ukrainian Orthodox Church in Vancouver (December 19, 1975). Reverend Peter Bublyk (left) is cutting a piece of Paul's hair, a Ukrainian tradition, as his godmother Lynn Zidek looks on

13.6 Post-publication Developments

I learned about many recent advances in ocean dynamics when writing the book, and as a spinoff, I started several new areas of research. After the book was finished, I wrote papers on the stability of eastern boundary currents [5, 6], nonlinear wave-wave interactions [7, 8], and ocean-climate variability and fish populations [9]. I also used the book for teaching courses on ocean waves at the Naval Postgraduate School, Monterey, CA (summer 1980) and the Swiss Federal Institute of Technology, Zurich (1982–83).

In the mid 1980s, Elsevier invited us to prepare a second edition of our book. Since I moved to McGill University in 1986 this was logistically quite a challenge. At the time, I was also leading a multidisciplinary research project [10] on the impact of climate variability on salmon migration in the Northeast Pacific (Project MOIST: Meteorological and Oceanographic Influences on Sockeye Tracks). However, over the course of two years, Paul and I worked on a revised version when I made several short visits to Vancouver. On nearly every page, we improved the presentation and introduced new research results. However, Elsevier was not interested in this kind of revision, as it would have involved typesetting a new manuscript. All Elsevier wanted was one long chapter of new results and observations which could be added to the end of the first edition. As educators and researchers, we did not find this

Photo 13.2 Top: On holidays with the LeBlond family at Sakanaw Lake (Sechelt Peninsula, BC), July 1978. Left to right: Philippe, Paul and Anne LeBlond, and my son Paul (age 3), daughter Claire (age 5 months) and me. Bottom: Mary Mysak holding Claire while overlooking strong tidal currents near the Sechelt Peninsula, BC

academically acceptable. Thus, we abandoned the revision project in 1988. As the consequence, it is hard to find an unused copy of the 1978 or 1980 version of *Waves in the Ocean,* but I was told a few years ago that Amazon could sell you a copy for about US $1000.

After I moved to Montreal, Paul and I kept in close touch, and we often shared fine dinners and wines at conferences. At age 57, Paul surprised many of us by taking early retirement in 1996, after three decades as a much-loved teacher and distinguished researcher at UBC. He and his second wife, artist Annette Shaw, moved to Galiano Island, BC in 1990, where he became active in many local historical and cultural organizations. While on Galiano, Paul also pursued his controversial studies of unidentified marine animals (sea monsters) such as "Cadborosauros" possibly sighted off the BC coast. In addition, he co-founded (with Edward Bousfield) the International Society of Cryptozoology.

We last met in 2015 at the annual meeting of our national academy, the Royal Society of Canada, in Victoria, where he was honored with a Life Membership. Paul was jovial as ever, and it was a joy to catch up on our news. Sadly, he died in February 2020 at age 81 (Photo 13.3), when I happened to be traveling in New Zealand. In his obituary, it was noted that Paul "was generous, practical, wise, funny, and expressed his zest for life in all he did." As an example of his generosity, he helped me move households twice in Vancouver, even though he still had young children at home. I think of Paul often and am very grateful for the friendship and collaboration we had over many decades.

SATURDAY, MARCH 14, 2020 | THE GLOBE AND MAIL

DEATHS

PAUL HENRI LEBLOND
December 30, 1938
February 8, 2020

Photo 13.3 Paul Henri LeBlond, from his obituary in The Globe and Mail, March 14, 2020. Photograph courtesy of Annette Shaw

Waves in the Ocean is the only book that I've written so far, as author or co-author. Remarkably, to the best of my knowledge, there have been no other books published in the past few decades that cover the recent advances in all classes of ocean waves. I feel very happy and content that over the past five decades, *Waves in the Ocean* has been so widely used by researchers, students, and professors.

13.7 Coda

After completing his PhD at UBC in 1978, Dan Wright went on to have an outstanding career as a research scientist at the Bedford Institute of Oceanography in Halifax, NS. In particular, he played a major role in developing a new thermodynamic equation of state for sea water (TEOS-10; see Chap. 22). He also co-supervised graduate students at Dalhousie University and presented thought-provoking papers at oceanographic conferences. In 1989, he spent a sabbatical at McGill University to broaden his research interests. While in Montreal he participated in the development and application of innovative climate models of reduced complexity [11, 12]. We often met at conferences and discussed his recent work in ocean and climate dynamics. Sadly, he died suddenly in 2010 when he was in his late fifties, which was a huge loss to Canadian oceanography. To honor his memory, the Canadian Meteorological and Oceanographic Society established a Graduate Scholarship in his name.

After returning to Laurentian University in 1976, Len Todd continued teaching mathematics and research in fluid dynamics. We met a few times at Canadian applied mathematics conferences in the 1980s, but we lost track of each other shortly thereafter. I assume he retired from Laurentian some time ago.

Rick Thomson and his wife Irma have two daughters born in the same years as my two children (1975 and 1978). Mary and I were thrilled that Rick agreed to be the godfather of our daughter Claire (Photo 13.4). In the 1970s and 1980s, they lived in a secluded forested property at the north end of the Saanich Peninsula, and we spent many weekends visiting them and taking hikes along the coastal ocean trails. We have kept closely in touch ever since.

Rick is about five years younger than me and continues to do research at the Institute of Ocean Sciences (IOS) on Vancouver Island. As well as being a prolific author of research articles, he has written several books on oceanography and data analysis methods. Early in his career, Rick wrote a very popular book on physical and geological oceanography used widely by west coast boaters, fishers, kayakers, and guides [13]. He also co-authored a reference book with Bill Emery on ocean data analysis [14]. Emery, a junior colleague of mine at UBC in the 1970s and 1980s, was very interested in remote sensing data and time series analysis [15, 16]. The Emery-Thomson (now Thomson-Emery) book and its subsequent second and third editions published in 2001 and 2014, respectively have been used by many students, researchers, and professors worldwide. My well-thumbed copy of the first edition has been borrowed by several of my former graduate students, and the book's pictures of oceanographic measurement equipment were frequently used in my introductory

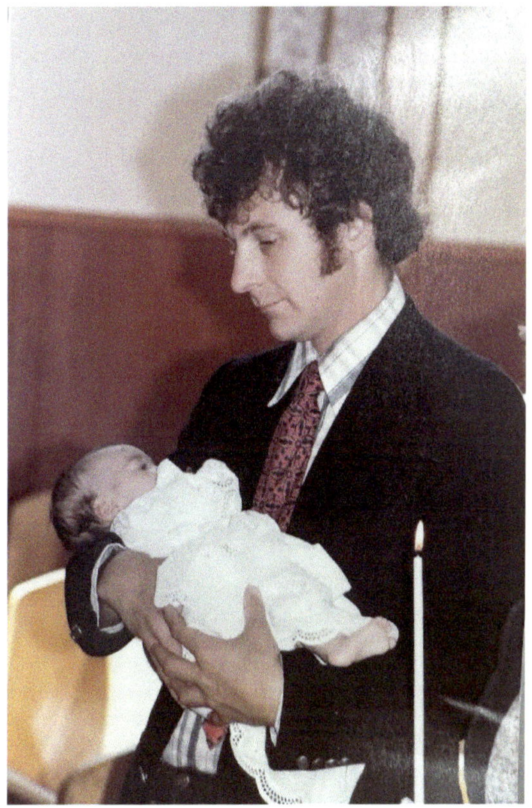

Photo 13.4 Rick Thomson holding my daughter Claire Anastasia Mysak at her christening on May 23, 1978, when she was three months old

ocean science course at McGill. Recently I learned that he and Emery have published a fourth edition of their book [17]. I was delighted to hear this news as the book will contain many new techniques used in ocean research today, such as machine learning and neural networks. After the publication of the fourth edition, Rick did officially retire from IOS in 2024 but continues to do research.

Acknowledgements I thank Janet Boeckh for reviewing the first draft of this memoir and Olivia Marino for assistance in inserting the figures in the manuscript. Josef Schmidt and Rick Thomson provided helpful comments on a later version of this chapter. I am grateful to Annette Shaw for giving me permission to use the photograph of Paul LeBlond that appears in Photo 13.3.

References

1. Phillips OM (1966) The dynamics of the upper ocean. Cambridge Univ Press, London
2. Turner JS (1973) Buoyancy effects in fluids. Cambridge Univ Press, London

3. Gossard EE, Hooke WE (1975) Waves in the atmosphere: atmospheric infrasound and gravity waves—their generation and propagation. Elsevier Scientific Publishing Co, Amsterdam
4. LeBlond PH, Mysak LA (1978) Waves in the Ocean. Elsevier Scientific Publishing Co, Amsterdam
5. Mysak LA (1977) On the stability of the California undercurrent off Vancouver island. J Phys Oceanogr 7:904–917
6. Mysak LA, Schott F (1977) Evidence for baroclinic instability of the Norwegian Current. J Geophys Res 82:2087–2095
7. Hsieh WW, Mysak LA (1980) Resonant interactions between shelf waves, with application to the Oregon shelf. J Phys Oceanogr 10:1729–1741
8. Mysak LA (1978) Resonant interactions between topographic planetary waves in a continuously stratified fluid. J Fluid Mech 84:769–793
9. Mysak LA, Hsieh WW, Parsons TR (1982) On the relationship between interannual baroclinic waves and fish populations in the Northeast Pacific. Biol Oceanogr 2:63–103
10. Mysak LA, Groot C, Hamilton K (1986) A study of climate and fisheries: interannual variability in the northeast Pacific Ocean and its influence on homing migration of sockeye salmon. Climatological Bull 20:26–35
11. Stocker TF, Wright DG, Mysak LA (1992) A zonally averaged, coupled ocean-atmosphere model for paleoclimate studies. J Climate 5:773–797
12. Wright DG, Stocker TF, Mysak LA (1990) A note on quaternary climate modelling using boolean delay equations. Clim Dyn 4:263–267
13. Thomson RE (1981) Oceanography of the British Columbia coast. Can Special Publ Fisheries Aquatic Sci 56
14. Emery WJ, Thomson RE (1998) Data analysis methods in physical oceanography (First Edition). Pergamon (Elsevier Science Inc., New York)
15. Emery WJ, Mysak LA (1980) Dynamical interpretation of satellite sensed thermal features off Vancouver Island. J Phys Oceanogr 10:961–970
16. Mysak LA, LeBlond PH, Emery WJ (1979) Trench waves. J Phys Oceanogr 9:1001–1013
17. Thomson RE, Emery WJ (2024) Data analysis methods in physical oceanography, Fourth Edition. Elsevier, Waltham, MA, USA

Chapter 14
A Magical Year in Zürich

Abstract The 1982–83 sabbatical in Zürich was very much a happy family affair. Mary and I learned German for the first time, the children went to a private French school (and learned elementary German too), we hiked scenic mountain trails, skied in the Alps and toured five other countries. A highlight for the children was an overnight train journey from Paris to Florence. We also enjoyed hosting the visits of many family members.

As well as developing a new theory for long-period topographic waves in the Swiss lakes while serving as a visiting professor at ETHZ, I gave seminars in Italy, France, Israel, Germany, and England. This outreach helped me develop a wide scientific network in Europe and beyond, which I still benefit from four decades later.

14.1 Why Zürich?

After the birth of our children, Mary urged me to take my next sabbatical in Europe by the ocean. We spent the last one in 1977 in Boulder, Colorado, where I was a visiting scientist in the ocean dynamics division of NCAR. Mary loved to travel and experience new countries and languages, and she thought it would be good for the children to live in another culture. My plan therefore was to find an institute in some European city where I could continue my oceanography research. So, how did we end up in Zürich for the year 1982–83? There are no oceans in Switzerland. Just lots of lakes and mountains.

Our Vancouver friends Kathy (originally from Switzerland) and Andrew Glass (Canadian) encouraged us to consider Switzerland for a sabbatical. Their two children, Christopher and Catriona, were born in the same months and years as Paul and Claire (June 1975 and February 1978). In 1979, Andrew took a job as a research physicist with the Swiss engineering firm Landis and Gyr in Zug, and in the following year we joined them for a family holiday to Switzerland. We loved seeing the mountains and lakes around Zug and nearby Zürich. Andrew arranged for me to give two seminars on ocean waves and current variability in the hydrology division of ETHZ, short for Eidgenössische Technische Hochschule Zürich. World-renowned for its

L. Mysak, *Adventures in Climate Science, Ocean Waves, and the Flute*, Springer Biographies, https://doi.org/10.1007/978-3-032-19848-8_14

research in the physical sciences and engineering, ETHZ is sometimes referred to as the 'MIT' of Switzerland. After my seminars, researcher Wilfred Horn showed me data on current and temperature measurements in the lakes of Zürich and Lugano (in the Italian part of Switzerland). "As you can see, Professor Mysak, some of the currents have oscillations with a period of several days, very much like what you described in your seminars," said Horn. "Maybe you have a theory to explain them?".

That was it. I would become a physical limnologist for the year 1982–83 trying to model and understand the sub-inertial (several-day period) current fluctuations in the Swiss lakes. The sabbatical would also be a big family adventure. We would have to find schools for the children, search for a furnished apartment, buy a car, and rent out our Vancouver home. In addition, to help our getting about in Switzerland, Mary and I studied German at UBC.

14.2 Settling into Zürich and ETHZ

Finding a three-bedroom apartment for the full year was hard. To this day, Paul remembers that we had to stay in a hotel for the first two weeks in September until the apartment was available. He loved the croissants and hot chocolate for breakfast there, and in the apartment, he can still picture the vines and grass outside his bedroom window that faced the Zürichberg, the small mountain to the east of the city of Zürich. Every morning at 8 am, Mary would drive the children to a private French school on top of the Zürichberg after dropping me off at the hydrology lab at ETHZ in the center of Zürich. Since Paul had started grade one in a French immersion program in Vancouver, we wanted him to continue his schooling in French. Fortunately, Claire was able to enroll in a "maternelle" (pre-school) program in the same school. On a few sunny mornings in the early fall, Mary would pick me up at ETHZ and we would go hiking in the spectacular mountains around Zürich (Photo 14.1).

I was a visiting professor at ETHZ on half salary and therefore offered a fall semester course on ocean waves. Dr. Kolumban, or Koli, Hutter was the director of the limnology group in ETHZ's hydrology lab and became my host there. He was renowned as a theoretical glaciologist [1] but also did research in many areas of fluid mechanics, including physical limnology, the study of waves, and currents in lakes. He also wanted to learn about ocean waves, and every week he attended my lectures on this topic. The lectures followed the early chapters of the book I had recently co-authored with Paul LeBlond [2].

14.3 The Three-Day Wave

During summer 1979, Hutter collaborated with geophysicists in Lugano to make measurements of wind, water temperature and the currents in the intermontane Lake of Lugano, to study its response to wind forcing. An analysis of these data revealed

Photo 14.1 Mary and me at the summit of Mount Klein Ibrig in the Swiss Alps near Zürich (October 1982)

an intriguing three-day oscillation in the temperature and current records. This was quite distinct from the more familiar wind-induced internal gravity wave oscillations (known as seiches) with periods of 24 h or less ([2], page 143). At the beginning of my sabbatical, Koli posed a challenge. "Lawrence, can you find a theory for this sub-inertial oscillation?" he asked. Subsequently, I developed a new theory for topographic waves [3] in a rotating fluid which in the northern hemisphere propagate around an elongated (elliptical) lake in an anticlockwise direction (Photo 14.2). In such a basin, the phase speed for a given wave mode traveling along an elliptical contour is variable, with its speed a maximum at the sides of the lake and a minimum at the ends of the lake. This is like the speed of a racing car around an elongated track.

When combining this new wave model with a theory for the vertical motion of the thermocline in a rotating stratified basin with steep sides [4], Koli and I derived a set of equations for the sub-inertial dynamics of the Lake of Lugano [5]. When we inserted into these equations the parameters that characterize the physical properties of this lake, we showed that sub-inertial currents in the model are surface-intensified,

Photo 14.2 Me working at the blackboard in my office at ETHZ, trying to find a mathematical description of the bottom topography of the Lake of Lugano, Switzerland. Note that the depth contours of the lake are approximately elliptical (December 1982)

in agreement with the observations [5]. Furthermore, following a several-day wind gust in the southern part of the lake, a free, fundamental mode elliptical topographic wave with a three-day period is generated in the model. Such a wave is observed to propagate around the lake four or more times before dying out ([5], Fig. 17). We argued [5] that this represents a type of natural oscillation in the Lake of Lugano.

The existence of topographic waves (also known as rotational or vorticity modes, see [2]) in the Lake of Lugano is surprising, considering the small horizontal dimensions of this alpine lake (approximately 17 × 1.5 km). At this scale, the Coriolis (rotational) effect is thought to be negligible. In a rotating stratified fluid, bottom-trapped topographic waves can exist [6], but these generally occur on horizontal scales of hundreds of kilometers. Since papers [3, 5] have 80 citations in total, it is conceivable that the wave theory we developed at ETHZ during my sabbatical may have been applied to other intermontane Lakes. I know for certain that Thomas Stocker, one of Dr. Hutter's PhD students (and later my postdoc at McGill), was motivated by [3] to investigate other forms of topographic waves in Swiss lakes for his doctoral thesis [7].

14.4 A Year of Enlightened Reading

During our many lunches together, Koli and I often talked about the lives and work of famous scientists and mathematicians who had spent time at ETHZ (e.g., Albert Einstein and Wolfgang Pauli) and other European universities (e.g., David Hilbert, Richard Courant, Neils Bohr, Max Born, Werner Heisenberg, Fred Hoyle, Lev Landau). The ETHZ library has many excellent biographies of these and other world-renowned scientists; I read over 15 of them during my sabbatical. I was often inspired by their work ethic, brilliance, social interactions, and their love of good food. Many enjoyed taking long walks and having scientific discussions in the forests, which prompted me to draft a short note for the journal Nature [8] during an extended coffee break while strolling in the Zürichberg Forest. This three-page paper summarized our results on the three-day wave in the Lake of Lugano.

I especially enjoyed reading Constance Reid's biography [9] of the mathematician David Hilbert (a professor at Göttingen University), and a follow up biography [10] of his famous student Richard Courant who spent the second half of his career at New York University, where he founded a Graduate School of Mathematical Sciences (now the Courant Institute). Hilbert was a versatile mathematician who worked in several areas throughout his lifetime. Every five to ten years he would dive into a new field, prove a few significant theorems, and then move on to another field. He supervised nearly 70 PhD students. Looking back on my 50 years as an academic, I have tried to emulate this approach with my own research. However, rather than prove theorems in different fields of mathematics, I have worked on specific applied mathematics problems in ocean and climate dynamics or analyzed large data sets depicting various climate phenomena.

Courant, together with Max Born, had established a world-renowned center for mathematics and physics in Göttingen University in the 1920s and early 1930s, after which when it was no longer possible for them, as Jews, to work in Germany. Prior to this period, Courant had an excellent apprenticeship as an "assistant" to Hilbert before heading the Mathematics Institute in this university. Courant loved to have many young students and postdocs around him, and he and his wife enjoyed hosting musical evenings and parties at their home. He continued in this style at New York University (NYU) in 1934 where he also did a lot of book writing, teaching and applied work with the US Office of Naval Research. The NYU's Institute of Mathematical Sciences was later named in his honor in the 1960s. In contrast to Hilbert, Courant liked to work on concrete problems. Donald Ludwig, who did research in mathematical epidemiology, was one of the last PhD students of Courant at New York University. He joined the mathematics faculty at UBC in the mid-1970s, and we became close colleagues and friends as applied mathematicians and musicians. Donald was an excellent cellist, and we often played chamber music together. Throughout my career, it has been great to meet many mathematical colleagues who are also talented musicians. Today, David Holland, one of my former PhD students at McGill, is a highly regarded professor of climate and ice dynamics at the Courant Institute in New York.

14.5 Family Travels and Visitors

Every school in Switzerland has a scheduled ski-holiday week. For Paul and Claire this turned out to be the last week of January. They and Mary had never skied before, and I had only spent a few winters skiing in Edmonton's North Saskatchewan river valley. Nevertheless, the Glass family enticed us to join them for a week in Sils Maria, a ski resort near St. Moritz. We rented downhill ski equipment for the week and took lessons every day on a beginner's hill, with a tow rope to get us to the top. The lessons were excellent, although Mary and I felt out of place among the young kids who were just starting out. By the end of the week, we could slowly slalom down this small hill. During the week, the children also took time to play in the snow (Photo 14.3). On the last day of our holiday, Paul and I skied down the main mountain in Sils Maria, which thrilled us both. I have never done any downhill skiing since. Today the whole family prefers cross-country skiing on forest trails.

The next month we did a major train trip to Paris, Florence, Lugano and back to Zürich. Shortly after we arrived in Switzerland, Paul kept asking, "Dad, when can we take an overnight train ride?" Since I was invited to give a set of lectures at the Physical Oceanography Lab in Paris and then wanted to visit my mathematician cousin Allan Trojan who was on sabbatical in Florence, I agreed that this would be

Photo 14.3 On a ski holiday at Sils Maria in the Swiss Alps. L to R: Christopher Glass, age 7 (family friend), my children Paul, age 7 and Claire, age 4 making a snow fort (January 1983)

the best time for this adventure. While Mary and I most enjoyed the culinary and visual delights of Paris (sidewalk cafes, art galleries and the Eiffel Tower), Claire was thrilled to get a birthday gift of roller skates in this city. A memorable experience for both children was the overnight train ride to Florence. Sleeping in the upper bunk as the train clicked along to Florence was Paul's biggest thrill of the trip.

In April, we visited Israel, thanks to an invitation from Mary's friends Mikael and Chava Katz whom she met in the 1960s while working on a kibbutz. As the Katz family lived in Rehovot, I arranged to give a seminar on coastal current dynamics at the famed Weizmann Institute in this city. After a few days visiting in Rehovot we rented a car to do a week-long whirlwind tour of Israel, visiting Jerusalem, Masada, the Dead Sea, and Haifa, where I gave an invited talk at the Technion. Two weeks was far too short for a visit to a country with so much history. However, the visit there did have a huge impact on my future scientific career, as described in Chap. 15. In Israel, I learned how a climate event (El Niño) in the distant Pacific could strongly affect the rainfall and fruit trees in the middle east.

Many of our family members visited us in Zürich. Mary's mother from London spent Christmas with us; she and Paul loved sharing a bag of freshly roasted chestnuts while sitting in the town square. My sister Helen came from Vancouver for a week in April and then took a holiday in the Greek Islands. In May, my father came from Vancouver for a week (Photo 14.4) and later did a tour of Austria. As a lover of Mozart, he couldn't wait to hear the master's music in the concert halls of Vienna. He even learned some rudimentary German to help with his travels in Austria.

14.6 Scientific Outreach

Visiting and giving seminars and conference papers in 14 different universities and research centers in six countries (England, France, Germany, Israel, Italy, and Switzerland) during my sabbatical was a very enriching experience. With the publication of *Waves in the Ocean* [2] in 1978 followed by my many research papers on half a dozen oceanographic topics (including climate-fish interactions [11]), my name was now well known in the European oceanographic community. Hence, I was a welcome seminar speaker. In December, I attended a NATO symposium/workshop on fish migration in southern Italy, where I was lucky to meet Kees Groot from the Pacific Biological Station (PBS) in Nanaimo, BC. He was interested to learn about the role of interannual oceanic variability (in currents and temperature) in the migration patterns of Pacific salmon, a topic I had just started to work on [11]. When I returned to Vancouver later in the fall, I met with Kees at PBS and shortly after we were successful in getting an NSERC Strategic Grant for a multi-disciplinary research project called MOIST: Meteorological and Oceanographic Influences on Sockeye Tracks [12]. Further information on this project is given in Chap. 15.

Photo 14.4 Claire, age 5, my dad, and Paul (nearly 8) beside the Lake of Zürich (May 1983)

In July, when Mary and the children were visiting family and friends in England, I was a guest lecturer at a summer school on Lake Hydrodynamics in Udine, Italy. Here I was excited to be able to publicize our new work on elliptical topographic (ET) waves in the Lake of Lugano [3, 5] as well as give introductory lectures on nonlinear internal waves [13]. Many colleagues teased us at the similarity of the name of our new waves with the title of the 1982 movie "ET: Extra-Terrestrial." Attendee Clifford Mortimer FRS, a father figure in physical limnology from Michigan (see his references listed in 8), was especially intrigued by the new model of ET waves [3].

At the end of August, I gave three papers at the IUGG General Assembly in Hamburg. Here I heard about the Pacific North American (PNA) teleconnection pattern in the atmosphere from Jerome Namais, an ocean-climate expert in California. After my talk on "Baroclinic Waves and Fish?" he jumped up in the audience. "It is well known that interannual oceanic temperature anomalies in the tropical Pacific will affect coastal North Pacific Ocean temperatures via the PNA pattern much more efficiently than baroclinic coastal trapped waves [11]," he said. After thinking about his comments, I ended up agreeing with him.

14.7 Epilogue

Overall, the year in Zürich was a happy time for the whole family (Photo 14.5). Mary became the godmother of Annabel, a daughter born to Andrew and Kathy Glass in fall 1983, which provided us with a good excuse to visit Zürich in future years. Mary and I were glad to return there for a six-month visit in 2000–2001, this time without the children. During this visit, I was a visiting professor in the Climate Institute of ETHZ since my research now focused on climate dynamics after I had moved to McGill in 1986. I gave a short course on ocean circulation and its role in climate at this institute.

Photo 14.5 Mary, Paul, age 8, Claire, age 5 [3] and me enjoying a sunny day at the water fountain in front of our Zürich apartment (July 1983)

We always liked the efficiency and cleanliness of the Swiss trains, as well as the beauty of the countryside. Every weekend Mary and I took extensive walks in the mountains. I was always amazed that at most train stops in small mountain villages, we could hop on a postbus to take us high into the mountains to a "Wanderweg" (walking trail), which inevitably led to a restaurant with strong coffee and outstanding views of the Alps. In March 2001 we moved to Bologna for the second half of the sabbatical, where I was a guest scientist at the National Institute for Geophysics and Volcanology. While we enjoyed the tasty food and friendliness of the Italians in Bologna, we missed the country walks that we had in Switzerland.

14.8 Farewell to Kolumban Hutter

The last time Mary and I saw Professor Hutter was at his 70th birthday celebration at ETHZ on January 22, 2011. Koli and I were always amused by the fact that he was precisely one year younger than me. Until a couple of years ago, we exchanged birthday greetings in January. At this celebration many of his former PhD students and colleagues gave papers on a large variety of topics, showing his great versatility in past research. Hutter was a very conscientious supervisor who could seem very stern initially. But once he warmed up to you, he was your friend for life.

On December 26, 2024, Thomas Stocker sent me the sad news that Kolumban Hutter had died just before Christmas. He was 83. This news brought back a flood of memories of the happy times we had together, at lunch, coffee breaks, conferences and boat trips (Photo 14.6). I am amazed at how many books Koli had written or edited during his career. When I was putting the finishing touches on this chapter, I discovered in my McGill office Volume 1 of a book series on "*Physics of Lakes*" that Koli had co-authored [13]. I was most touched when I read, for the first time, that he dedicated this book series to distinguished limnologist Clifford Mortimer (1911–2010) and myself, as "mentors, teachers, and friends."

Photo 14.6 Kolumban Hutter (at left) and me in the limnology research boat on the Lake of Zürich, celebrating my farewell as a visiting professor at ETHZ (August 1983). Sadly, Kolumban passed away 20 December 2024, at age 83

Acknowledgements I thank Janet Boeckh for comments on the first draft of this memoir, Claire Mysak for inserting the photographs, Joseph Schmidt for his excellent critique, and Nicholas Acheson, Roussos Dimitrakopoulos and Linda Johnston for their input.

References

1. Hutter K (1983) Theoretical glaciology. D Riedel, Dordrecht
2. LeBlond PH, Mysak LA (1978) Waves in the ocean. Elsevier Scientific Publishing Co, Amsterdam
3. Mysak LA (1985) Elliptical topographic waves. Geophys Astrophys Fluid Dyn 31:93–135
4. Gratton Y (1983) Low-frequency vorticity waves over strong topography. PhD thesis, University of British Columbia, Vancouver
5. Mysak LA, Salvade G, Hutter K, Schweiller T (1985) Topographic waves in a stratified elliptical basin, with application to the Lake of Lugano. Phil Trans R Soc Lond A316:1–55
6. Rhines PB (1970) Edge-, bottom-, and Rossby waves in a rotating stratified fluid. Geophys Fluid Dyn 1:273–302
7. Stocker T (1987) Topographic waves: Eigenmodes and reflection in lakes and semi-infinite channels. Mitteilung Nr 93 der Versuchsanstalt für Wasserbau, Hydrologie und Glaziologie an der ETHZ
8. Mysak LA, Salvade G, Hutter K, Schweiller T (1983) Lake of Lugano and topographic waves. Nature 306:46–48
9. Reid C (1970) Hilbert. Springer-Verlag, New York

10. Reid C (1976) Courant in Göttingen and New York. The story of an improbable mathematician, Springer-Verlag, New York
11. Mysak LA, Hsieh WW, Parsons TR (1982). On the relationship between interannual baroclinic waves and fish populations in the Northeast Pacific. Biol Oceanogr 2:63–103
12. Mysak LA, Groot C, Hamilton K (1986) A study of climate and fisheries: interannual variability in the northeast Pacific Ocean and its influence on homing migration of sockeye salmon. Climatological Bull 20:26–35
13. Hutter K, Wang Y, Chubarenko IP (2011) Physics of lakes. In: Volume 1, Foundation of the mathematical and physical background. Springer, Heidelberg Dordrecht London New York

Chapter 15
El Niño and Me: My Move to McGill University

Abstract Our 1982–83 sabbatical in Zürich was a pivotal year in my scientific career. While visiting Israel and Germany on a lecture tour in spring 1983, I learned about the El Niño warming event of the century in the tropical Pacific Ocean. This natural climatic event really caught my attention as it had significant impacts on fisheries in the eastern Pacific and strongly affected the weather in North America and beyond, even in the Middle East. When I returned to UBC after my sabbatical, I started a research project on fish-climate interactions involving the Fraser River Sockeye salmon migration. This in turn inspired me to search for a new position where I could focus on climate research, which was not being supported by UBC.

In the summer of 1985, I was offered a professorship at McGill University to focus on climate research, and in August 1986, we moved to Montreal where my academic home was in the Department of Meteorology (now named Atmospheric and Oceanic Sciences). In a few years, I became the founding director of the multidisciplinary Centre for Climate and Global Change Research (C^2GCR), which brought together faculty and graduate students from several departments at McGill.

A sweet but mysterious fragrance hits our nostrils as we drive away from Tel Aviv airport in April 1983. "It's from the orange trees, now bursting with ripe fruit," my wife's friend Mikael says. "We'll see many of these trees around Rehovot, where I live. We've not had fruit crops like these for many years. It's probably because we've had a huge amount of rain this spring. And the deserts are blooming with bright red and yellow flowers."

We didn't think much further about these unusual conditions during our family visit to Israel. Instead, we focused on touring the historic sights of Jerusalem, Masada and the Dead Sea. Also, I was preparing presentations on theoretical oceanography to be given in Rehovot and Haifa.

After returning to Zürich, where I had been on sabbatical since September 1982 (Chap. 14), I continued developing mathematical models of varying current patterns in the Swiss lakes. In May 1983, I gave a seminar on this topic at the marine institute in Kiel, Germany. There Professor John Woods, a distinguished oceanographer from England, showed me recent maps of the sea surface temperature of the tropical Pacific.

L. Mysak, *Adventures in Climate Science, Ocean Waves, and the Flute*, Springer Biographies, https://doi.org/10.1007/978-3-032-19848-8_15

In a wide swath along the equator, the ocean temperatures were a remarkable 2 to 3 °C above the long-term average.

"This is the El Niño of the century," John said. "This huge pool of warm water generates long atmospheric waves which propagate around the globe at mid-latitudes. These dramatically alter the precipitation patterns in North America, Europe *and* the Middle East."

So, this is what probably produced the unusual crops and flowers in Israel. Clearly, I had to learn about El Niño, and I looked forward to doing so after returning in September 1983 to my academic home in the Mathematics Department at UBC, where teaching maths was my livelihood.

El Niño is a warming event in the tropical Pacific Ocean that occurs every three to seven years and lasts about twelve months. El Niño means "the Christ child" in Spanish and is used to describe these warming events since they start around Christmas. Alternating with El Niño is La Niña, a cooling event of the tropical Pacific Ocean.

In late 1983, back in Vancouver, I became even more intrigued about the 1982–83 El Niño when Cornelis (Kees) Groot, a fisheries scientist from Nanaimo, showed me the latest statistics on the salmon migration routes in the northeast Pacific. In fall 1983, 85% of the Fraser River Sockeye salmon came home to spawn via the northeast end of Vancouver Island (Fig. 15.1). The remainder returned to the Fraser River via Juan de Fuca Strait, to the south of Vancouver Island. In the past, except for a few years, returning Sockeye mostly took the southern route. The 1982–83 El Niño event caused unusually warm water west of Vancouver Island. Since salmon prefer cold water, the returning Sockeye were farther north than usual, close to Alaska, and hence came home via the northern route. This was good news for Canadian fishers since treaty agreements do not allow American fishers to catch Sockeye in Canadian waters east of Vancouver Island. To get a deeper understanding of the past migration patterns of the Sockeye and to predict their future ones, I formed a research team with Groot and meteorologist Kevin Hamilton and started Project Meteorological and Oceanographic Influences on Sockeye Tracks (MOIST), which received major support through a Strategic Grant from Natural Sciences and Engineering Research Council of Canada (NSERC) [1].

I was hooked. El Niño and its impacts became my research focus from that point on. While boning up on El Niño (EN), I learned about an atmospheric circulation pattern in the Pacific called the Southern Oscillation (SO) which is coupled to it to create a complex climate phenomenon known as ENSO. I wrote a review entitled "*El Niño, interannual variability and fisheries in the northeast Pacific Ocean*" about ENSO and El Niño [2]. It is now one of my most highly cited papers, which may sound strange since I knew nothing about ENSO before my 1982–83 sabbatical. In retrospect, it is amazing that an observation made during a family visit to Israel sparked a change in my own career.

In the early 1980s, natural climate variability (like ENSO) and global warming due to our massive burning of fossil fuels and land use change caught the attention of the press and the public eye. My PhD Harvard classmate, George Philander, wrote a colorful article about ENSO for *National Geographic*. I pinched his pictures for

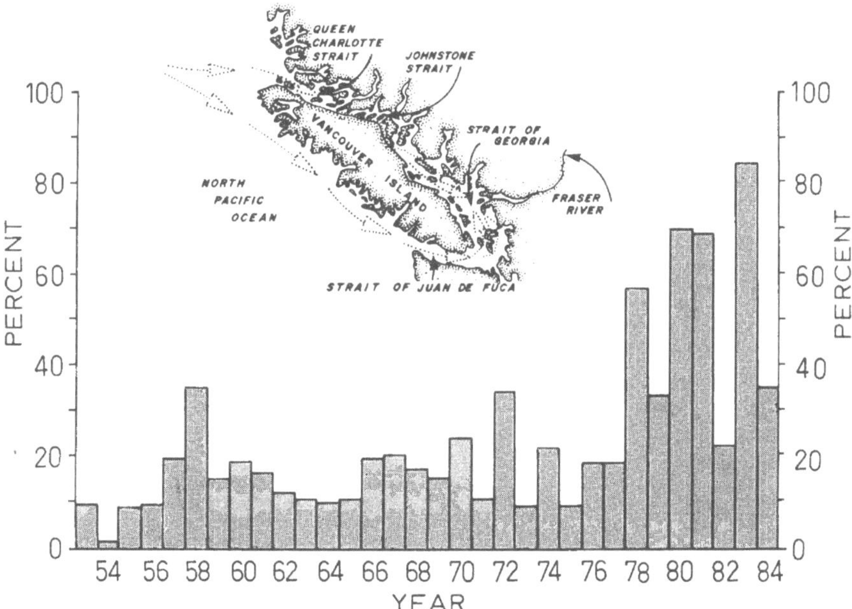

Fig. 15.1 Insert at top: The two migratory routes of adult sockeye salmon on their way to the Fraser River and other southern British Columbia and Washington State spawning grounds: the northern route through Johnstone Strait, and the southern route through the Strait of Juan de Fuca. Histogram: percentage of sockeye salmon approaching the Fraser River each year through the northern passage during 1953–84 (the so-called Johnstone Strait diversion) (From [2])

my public lectures about ENSO and fisheries. I also started thinking that I would like to work in a place where climate research was strongly supported. UBC was not likely to become active in this field. When approached by Environment Canada about sponsoring a professorship in climate research, the university administration declined to participate because of financial constraints at UBC. This reinforced my thoughts of a mid-career move. I was 44.

Sometime in 1984, I was invited to apply for an applied mathematics professorship at the University of New South Wales in Sydney, Australia. I was runner-up for this position. Ted Buchwald, the department head, phoned to say, "You certainly deserve the appointment, but I don't think you would move your family here. Instead, I'm offering the professorship to Roger Grimshaw, from Melbourne." I knew Roger well, and he certainly was an excellent candidate for the job. But this got me riled up. *Damn it, I said to myself, let's see if I'm marketable elsewhere.*

However, getting a senior position after two decades at UBC wouldn't be easy. In addition, Mary, who joined me in Vancouver in 1972 and gave birth to our two children there, was happily settled. By this time Mary was enjoying her studies in the humanities at UBC. "Can I do this in a new place?" she wondered.

Early in 1985, I sent my CV to colleagues in six attractive places where I might do oceans and climate research, namely, Honolulu, Seattle, Corvallis (Oregon), Zürich,

Hobart (Tasmania), and Montreal. This led to two interviews, the first in March at McGill University and the second in May at the National Oceanography Lab in Hobart. In June, McGill offered me a newly created Chair in Climate Research in the Department of Meteorology, where I was expected to initiate an internationally recognized climate research program. Shortly afterward, the Hobart Lab offered me the position of Chief Scientist to head up research on oceanography, climate and fisheries. With these two offers, it was clear that my career was going in a new direction and that I could soon be facing interesting challenges.

But I was in a quandary. I loved the interdisciplinary aspect of the job in Hobart. But this would be in a non-academic setting. On the other hand, at McGill I would be in a department for which I had no formal training (meteorology). However, McGill was known to attract excellent graduate students. But *would I miss teaching mathematics?* When word got out that I might leave Canada for Australia, colleagues pressed me to accept the McGill position. This I did in July 1985.

Mary knew she would miss Vancouver and the spectacular view of the ocean and snow- capped mountains from our bedroom window. But she understood my desire for this career change. We gathered the children in the family room to tell them of the upcoming move.

Claire, who was seven, burst into tears. "I won't have any friends in Montreal," she shouted. However, she took the news much better when she learned that there were very good gymnastic clubs in Montreal. Gym was her favorite sport.

Paul, on the other hand, took the news more pragmatically. "Dad, I'm glad you didn't choose Tasmania because I didn't want to have to learn a third language, *Tasmanian*," he said. Being 10 at the time, Paul already had spent several years in French immersion. He was confident that he could cope with French in Montreal.

Upon arriving in Montreal in August 1986, we quickly settled into our new home in Montreal West, a village within a metropolis. Mary soon made friends, helped edit the *Informer*, the local newsletter, and exhibited her photographs at the "Arts, etc.," event each fall. Mary studied photography at Dawson College in the late 1980s, and in 1992, enrolled at McGill, where she continued working part time towards a BA degree. In 1999, she was the proud holder of a first-class honors BA diploma in Asian Religions.

Our children were ultimately happy with our move to Montreal. They did well at school and in sports. Paul won a top athletic award upon graduation from Lower Canada College and went straight to the University of New Brunswick in 1992 to enroll in a forestry/engineering degree program. At age 17, Claire won the provincial championship on beam in gymnastics; she left home at 19 to attend Bishop's University (in Sherbrooke, QC), focusing on drama.

The move to McGill also provided me many new opportunities. With funding from McGill, NSERC, and the Quebec government, I founded the multi-faceted Centre for Climate and Global Change Research (C^2GCR) in 1990 (Chap. 17). This attracted highly motivated graduate students, as well as a steady stream of visitors from abroad whom I often entertained at the historic McGill Faculty Club [3]. I expanded my research interests to include the Arctic climate system [4–10], air-ocean interactions in the Atlantic [11–13], paleoclimates [14–23], modelling future global warming scenarios [24–26], and climate impacts [27–29]. I also developed interdisciplinary courses in climate and paleoclimate dynamics, which were popular with the students.

Although I officially retired from McGill in 2010 at age 70, I continued to supervise graduate students for nearly another decade. Nearly six decades on from 1967 when I started my academic career in Vancouver, I look back with fondness on the 80 graduate and postdoctoral students that I have supervised or co-supervised at UBC and McGill. They are my 'academic children'. Several of these are profiled in Chap. 24. Twenty-one are now professors in about a dozen countries. It is gratifying to know that they are furthering our understanding of ocean dynamics and its role in climate change. In addition, many of them are informing the public about the environmental dangers due to global warming such as rising sea levels, ocean acidification, extreme storms, and intense heat waves and forest fires. These students are also helping us learn about the mitigating actions that we can take to sustain our life on this planet.

15.1 Endnote

This chapter is an expanded version of my final assignment in the Thomas More Institute (TMI) Memoir Writing Course that I took in winter 2018. I am very grateful to Pauline Beauchamp and Karen Nesbitt, the course instructors, for their feedback and suggestions for the improvement of this memoir.

References

1. Mysak LA, Groot C, Hamilton K (1986) A study of climate and fisheries: interannual variability in the northeast Pacific Ocean and its influence on homing migration of sockeye salmon. Climatological Bull 20:26–35
2. Mysak LA (1986) El Niño, interannual variability and fisheries in the northeast Pacific Ocean. Can J Fish Aquat Sci 43:1464–1497
3. Stewart NF (2024) From the Baumgarten house to the McGill faculty club. Marking the centenary of the faculty club 1924–2024. Accent Impression, Montreal QC
4. Holland DM, Mysak LA, Manak DK, Oberhuber JM (1993) Sensitivity study of a dynamic thermodynamic sea-ice model. J Geophys Res 98:2561–2586
5. Mysak LA, Manak DK (1989) Arctic sea-ice extent and anomalies 1953–84. Atmos-Ocean 27:376–405
6. Mysak LA, Venegas SA (1998) Decadal climate oscillations in the Arctic: a new feedback loop for atmosphere-ice-ocean interactions. Geophys Res Lett 25:3607–3610
7. Mysak LA, Manak DK, Marsden RF (1990) Sea-ice anomalies observed in the Greenland and Labrador seas during 1901–1984 and their relation to an interdecadal Arctic climate cycle. Clim Dyn 5:111–133
8. Tremblay L-B, Mysak LA (1997) Modeling sea ice as a granular material, including the dilatancy effect. J Phys Oceanogr 27:2342–2360
9. Venagas SA, Mysak LA (2000) Is there a dominant timescale of natural climate variability in the Arctic? J Climate 13:3412–3434

10. Wang J, Mysak LA, Ingram RG (1994) Interannual variability of sea-ice cover in Hudson Bay, Baffin Bay and the Labrador Sea. Atmos-Ocean 32:421–447
11. Peng S, Mysak LA, Ritchie H, Derome J, Dugas B (1995) On the differences between early and midwinter atmospheric responses to sea surface temperature anomalies in the northwest Atlantic. J Climate 8:137–157
12. Venegas SA, Mysak LA, Straub DN (1996) Evidence for interannual and interdecadal climate variability in the South Atlantic. Geophys Res Lett 23:2673–2676
13. Venegas SA, Mysak LA, Straub DN (1997) Atmosphere-ocean coupled variability in the South Atlantic. J Climate 10:2904–2920
14. Antico A, Marchal O, Mysak LA (2010) Time-dependent response of a zonally-averaged ocean-atmosphere-sea ice model to Milankovitch forcing. Clim Dyn 34:763–779. https://doi.org/10.1007/s00382-010-0790-6
15. Brault M-O, Mysak LA, Matthews HD, Simmons CT (2013) Assessing the impact of late Pleistocene megafaunal extinctions on global vegetation and climate. Climate of the Past 9:1761–1771. https://doi.org/10.5194/cp-9-1761-2013
16. Carozza DA, Mysak LA, Schmidt GA (2011) Methane and environmental change during the Palaeocene-Eocene thermal maximum (PETM): modeling the PETM onset as a two-stage event. Geophys Res Lett 38:L05702. https://doi.org/10.1029/2010GL046038
17. Marson JM, Mysak LA, Mata MM, Wainer I (2016) Evolution of the deep Atlantic water masses since the last glacial maximum based on a transient run of NCAR-CCSM3. Clim Dyn 47:865–877. https://doi.org/10.1007/s00382-015-2876-7
18. Mysak LA (2008) Glacial inceptions: past and future. Atmos-Ocean 46:317–341. https://doi.org/10.3137/ao.460303
19. Schmidt GA, Mysak LA (1996) Can increased poleward oceanic heat flux explain the warm Cretaceous climate? Paleoceanography 11:579–593
20. Sedláček J, Mysak LA (2009) A model study of the Little Ice Age and beyond: changes in ocean heat content, hydrography, and circulation since 1500. Clim Dyn 33:361–475. https://doi.org/10.1007/s00382-008-0503-6
21. Stocker TF, Mysak LA (1992) Climatic fluctuations on the century time scale: a review of high-resolution proxy data and possible mechanisms. Clim Change 20:227–250
22. Stocker TF, Wright DG, Mysak LA (1992) A zonally averaged, coupled ocean-atmosphere model for paleoclimate studies. J Climate 5:773–797
23. Wang Z, Mysak LA (2006) Glacial abrupt climate changes and Dansgaard-Oeschger oscillations in a coupled climate model. Paleoceanography 21:PA2001, https://doi.org/10.1029/2005PA001238
24. Brault M-O, Matthews HD, Mysak LA (2017) The importance of terrestrial weathering changes in multimillennial recovery of the global carbon cycle: a two-dimensional perspective. Earth Syst Dyn 8:455–475. https://doi.org/10.5194/esd-8-455-2017
25. Cochelin A-SB, Mysak LA, Wang Z (2006) Simulation of long-term future climate changes with the Green McGill paleoclimate model: the next glacial inception. Clim Change 79(381–410):8. https://doi.org/10.1007/s10584-006-9099-1
26. Petoukhov V, Claussen M, Berger A, Crucifix M, Eby M, Eliseev AV, Fichefet T, Ganopolski A, Goosse H, Kamenkovich I, Mokhov I, Montoya M, Mysak LA, Sokolov A, Stone P, Wang Z, Weaver AJ (2005) EMIC Intercomparison Project (EMIP–CO_2): comparative analysis of EMIC simulations of climate, and of equilibrium and transient responses to atmospheric CO_2 doubling. Clim Dyn 25:363–385. https://doi.org/10.1007/s00382-005-0042-3
27. Damyanov NN, Matthews HD, Mysak LA (2012) Observed decreases in the Canadian outdoor skating season due to recent winter warming. Environ Res Lett 7:8
28. Hsieh WW, Lee WG, Mysak LA (1991) Using a numerical model of the northeast Pacific Ocean to study the interannual variability of the Fraser River sockeye salmon (Oncorhynchus nerka). Can J Fish Aquat Sci 48:623–630
29. Simmons CT, Mysak LA (2012) Stained glass and climate change: how are they connected? Atmos-Ocean 50(2):219–240

Chapter 16
On Becoming a "Fellow"

Abstract In April 1986, I learned of my election as a Fellow of the Royal Society of Canada (FRSC), often considered the highest Canadian honor for an academic in any field of science, the humanities, and the social sciences. My then wife Mary and I went to Winnipeg, Manitoba for the induction ceremony, which was held in June at the University of Manitoba. Here we met newly and previously elected Fellows from across Canada, including Don Betts, one of my favorite undergraduate physics professors at the University of Alberta. After attending various business meetings of the RSC, I vowed that after settling into McGill in September 1986 (Chap. 15) I would get involved in the activities of the academy. Little did I know that in less than a decade I would be presiding over the Academy of Science, the largest of the three academies that comprise the RSC (Chap. 19).

"Did you hear the good news about Ernie Kanasewich?" Dad asked over the phone in April 1976. He was calling from Edmonton to tell me that Ernie had just been elected a Fellow of the Royal Society of Canada. Ernie was a well-known senior professor of geophysics at the University of Alberta and a friend of the family. His parents were immigrants from Ukraine with little education, so this was considered a huge honor by members of Edmonton's Ukrainian community, who had a high regard for education and scholarship.

"Wow, that's fantastic," I replied. Like many colleagues and Canadians generally, I knew little about the Royal Society of Canada (RSC). But I did know that our new chair of the Department of Mathematics at UBC, Don Bures, had been elected FRSC in 1973, and this was one of the reasons he was selected for this position. Election to the RSC, Canada's national academy of the arts and sciences, is considered the highest honor for an academic.

Always being ambitious for me, Dad next asked, "So when do you think you will be nominated to the RSC?" This caught me by surprise. "Dad," I answered, "I have just been promoted to full professor at UBC, and, at the age of 36, I don't think that an FRSC is in the cards for me now."

L. Mysak, *Adventures in Climate Science, Ocean Waves, and the Flute*, Springer Biographies, https://doi.org/10.1007/978-3-032-19848-8_16

Much later, I learned that there was a lot of work in making an RSC nomination and often politics was involved in getting elected FRSC. In addition, most new Fellows were in their fifties or sixties when elected.

My own story about becoming a Fellow of the RSC began nearly a decade later. In the spring of 1985, when I had decided to leave UBC, I was in a quandary about whether to accept the offer of a senior research position (Chief Scientist) in oceanography with Commonwealth Scientific and Industrial Research Organization (CSIRO) in Tasmania, Australia, or a research chair in climate science at McGill, in Montreal. As an enticement for accepting the job in Tasmania, Angus McEwan, a Fellow of the Australian Academy of Science (FAA), and director of the CSIRO lab, said during my interview, "Of course in a few years, we will put you up for FAA."

On my return to Canada after the Australian interview, I was invited to give an applied mathematics lecture at the University of Toronto. There I happened to mention the possible FAA nomination to my colleague Keith Ranger FRSC, an applied mathematician who worked in fluid mechanics, and who was familiar with my own research accomplishments.

"Well," he remarked, "maybe it's time we put you up for FRSC. Please send me your CV." I did this when I returned to UBC, and then a few months later I was delighted to learn that he indeed submitted my nomination, which was co-sponsored by Jim Arthur FRSC, a mathematician from Toronto, and Tim Parsons FRSC, an oceanographer from UBC. Later I heard that Arthur was elected a Fellow of the Royal Society of London (FRS), which is considered the top honor for a UK scientist and any scientist in the British Commonwealth. Right after making the nomination, Keith went off to Australia for a sabbatical, and he left some of the paperwork to be completed by Tim Parsons. Tim grumbled to me later about having to do this extra work, but he did so in a good-natured way. I do believe he thought I deserved the honor.

In the meantime, I accepted the McGill offer in the summer of 1985 and started making plans with my family to move to Montreal in August 1986. In April, *the letter* from the RSC secretary arrived.

"Would you be willing to accept the honor of Fellowship in the Royal Society of Canada?" the secretary wrote. "If you are, please send us your consent and a cheque for the initiation fee and annual dues." I was both stunned and thrilled. In those days, a nomination often remained in the books for several years before one was elected. To be elected on the first go was not expected, especially at the age of 46. The request for the cheque immediately suggested that the Royal Society, a non-governmental organization, was not flush with cash.

Mary, my then wife, did not know about the nomination. When the letter arrived in the morning mail, I immediately invited her to the UBC Faculty Club for lunch. "I have some news to share with you," I said over the phone. Before going in for lunch,

Photo 16.1 Me at the induction as a new FRSC in Winnipeg, June 1986. The diagonal line across my name tag indicates a new fellow

we sat on a bench overlooking Howe Sound as she read the letter. I could see happy tears in her eyes when she got to the end. "Wonderful!" she said. "This makes up for the big disappointment you had in 1982, when Paul LeBlond was elected FRSC."

In 1978, Paul and I completed the 600-page treatise *Waves in the Ocean*, which was highly acclaimed and for which Paul and I were jointly awarded the President's Prize from the Canadian Meteorological and Oceanographic Society (Chap. 13). Thus, I was surprised and felt very deflated when in the spring of 1982 Paul's election to the RSC was announced. To Mary's credit, she knew from Paul's wife that he was about to be elected to the RSC but chose not to tell me. It took me a long while to get over this disappointment when the news finally broke about his election. I never knew who nominated Paul for FRSC, but it was clear that he had good friends in high places!

The induction of new Fellows for the class of 1986 was held in Winnipeg, Manitoba, as an add-on to the annual meeting of the Learned Societies which we could also attend (Photo 16.1). It was June and the city was infested with giant mosquitoes. Fortunately, most of the meetings and social events were held indoors. Mary came to the induction (Photo 16.2), but she had to leave Winnipeg early to attend the unexpected funeral of her mother in London, UK. I was very touched that she made the trip to Winnipeg to share this special event with me even though she was mourning the loss of her mother.

Photo 16.2 Mary Mysak, my only guest at the RSC meeting in Winnipeg, June 1986

To my great delight, Ernie and Elaine Kanasewich drove to Winnipeg from Edmonton to attend the RSC meeting. This was ten years after Ernie's own election to the RSC. Their warm congratulations meant a lot to me. I was also very happy to meet my former University of Alberta physics professor Don Betts FRSC, who was now Dean of Science at Dalhousie. He helped guide me through the different business meetings and social events, and he also explained the basic structure of the RSC. At that time, the Society was comprised of three academies: Académie des lettres et des science humaines, the Academy of Humanities and Social Sciences, and the Academy of Science/Académie des sciences, the largest of the three academies. I was elected as a member of the third academy, which was officially bilingual, unlike the other two academies. Don Betts encouraged me to get involved in the various committees of both the Academy of Science and the RSC, where I would interact with scientists and scholars from many fields. I felt inspired by this meeting and made a commitment to do this after I had settled into McGill.

One of the highlights of the science academy meeting was a large circular gathering where each of the 25 inductees had to give a two-minute "elevator talk" about their research and academic background. I highlighted my Ukrainian roots in a city which had a large Ukrainian community (Photo 16.3), and I also talked about the interdisciplinary nature of my work in applied mathematics, fisheries oceanography, and climate variability. At the induction, I was cited for "notable applications of mathematics to oceanography and ocean climate-fisheries interactions." My elevator talk caught the attention of Chris Barnes, then a member of the governing Council of the science academy; he later asked whether I would agree to also serve on the Council. Again, I managed to put this off until I had spent a year at McGill.

16.1 Epilogue

After four decades as a Fellow of the RSC, I still enjoy going to the annual meeting of the Society, which is now called the "Celebration of Excellence and Engagement." This lofty title is used to reflect the working motto of the RSC: "to promote and recognize excellence, and to serve society."

I did spend over a decade serving in various leadership roles of the RSC, culminating in my being president of the Academy of Science during 1993–96 (Chap. 19). This was followed by three years as past president and chair of the academy-wide selection committee for new Fellows. Shortly after my election, I nominated several colleagues for FRSC and have continued to do so. It has been a great pleasure to meet and work with many outstanding scholars and scientists from across the country, and several Fellows have become close friends.

Sadly, many of these have deceased, including Ernie Kanasewich, Don Betts, and Paul LeBlond, the latter in February 2020. It has been observed that Fellows of the RSC live several years longer than the average lifespan for Canadians. Jokingly, when I was president of the science academy, I used to tell new Fellows at their induction that "five years has just been added to your life." I hope this is still true.

Photo 16.3 Top: The Ukrainian Orthodox Cathedral in Winnipeg. Bottom: Me in front of the statue of Taras Shevchenko (1814–61), Ukraine's most famous poet who was a champion of the Ukrainian language and a promoter of Ukraine's independence

Acknowledgements Many thanks to Janet Boeckh for her eagle eye for noting redundancy, lack of clarity, and grammatical errors, and to Olivia Marino for assistance with the photographs. The insightful comments of Josef Schmidt are much appreciated.

Chapter 17
Creating a Climate Centre for McGill

Abstract In 1990, I proposed to the McGill senior administration that an interdisciplinary Centre for Climate and Global Change Research (C^2GCR) be established. The Centre would bring together 13 professors and their graduate students from four departments who did research into various aspects of climate variability and terrestrial global change. C^2GCR was supported by McGill, the Canadian Atmospheric Environment Service, NSERC and the Québec government, and it quickly became recognized nationally and internationally as a "go-to" place for research and graduate studies in these areas. After I served as founding director for six years, the Centre was admirably led by Nigel Roulet, Gail Chmura, Wayne Pollard, and Charles Lin. However, due to changes in provincial government funding and other scientific developments in Québec, the Centre disbanded in 2015, in its 25th year.

Shortly after I arrived at McGill University in the summer of 1986, Roddy Rogers, the Chair of the Meteorology Department, had a serious chat with me. He said "I want you to put climate research at McGill on the map. How you do this is up to you, but I envision a group of faculty members working together to establish a unit that would be the 'go-to' place in Canada for graduate studies and research in climate."

I knew this would be a challenging task. My previous position had been teaching applied mathematics and physical oceanography at UBC, and I had only recently started research in climate science. Prior to coming to McGill, I started a collaborative research project on the influence of interannual atmospheric and oceanic variability on sockeye salmon migration in the northeast Pacific [1]. At the same time, I had been learning about El Niño events in the tropical Pacific and how the oceans played an important role in climate variability and long-term change. However, at McGill I was joining a department of meteorologists who mainly studied clouds and weather. They had little experience in the latest research in natural climate variability (such as El Niño) and global warming due to the burning of fossil fuels and land-use change (anthropogenic forcing of climate change). My mandate was to work with colleagues from this and other McGill departments to develop new teaching and research programs in these fields. Rogers appointed me as director of a "Climate Research Group" (CRG).

L. Mysak, *Adventures in Climate Science, Ocean Waves, and the Flute*, Springer Biographies, https://doi.org/10.1007/978-3-032-19848-8_17

Photo 17.1 Charles Lin (left) and the author in 1987 in front of a blackboard describing the climate system: the atmosphere, oceans, cryosphere (sea ice and ice sheets), lithosphere (Earth's crust), and biosphere (Earth's living material)

In 1985, Rogers obtained substantial funding for climate studies at McGill via the Atmospheric Environment Service (AES)/NSERC Industrial Chair in Climate Research. The Chair funding paid the salary (for five years) for two professors, one senior (myself) and one junior (Charles Lin, who came from the University of Toronto) (Photo 17.1). The Chair funding also enabled us to hire a secretary, bring in seminar speakers, support a couple of postdoctoral fellows, publish a newsletter, and purchase new computing equipment. In addition, Bill Leggett, the then Dean of Science, and Gordon Maclachlan, the then Dean of Graduate Studies and Research, strongly encouraged us to carry out interdisciplinary research which would break down academic silos. In the beginning, the CRG brought together five faculty (three atmospheric scientists and two oceanographers) from the Meteorology Department and two faculty (physical geographers) from the Geography Department.

Starting in 1987, I encouraged the printing and circulation of technical reports that consisted of unpublished scientific articles by CRG faculty and their students, which were under review for journals, and maps of atmospheric and oceanic data used by the faculty in their climate research. These reports were listed in the quarterly CRG Newsletter that was mailed to colleagues around the world. In the newsletter, we also advertised our seminar speakers and the research activities of our members. In this manner, the CRG soon became known internationally. "We're on the map," Roddy used to say.

One of the fun things I enjoyed in my first couple of years at McGill was the organization of a full-year graduate climate course that was jointly taught by four members of the CRG: Charles Lin (atmospheric dynamics scientist), Kevin Hamilton (global upper-atmosphere circulation expert), John Lewis (physical geographer knowledgeable in atmosphere-land interactions), and myself (oceanographer and natural climate variability specialist). In this course, the students learned that climate science went far beyond studying the average of the weather in different latitudinal zones on the planet, a subject traditionally known as climatology. We taught them about various types of natural climate variability on timescales of years, decades, and longer. This involved exploring the nature of air-sea and air-ice-sea interactions and feedbacks, and atmosphere-land-vegetation interaction processes. We also showed the students evidence of the gradual warming of the planet due to the increasing concentration of carbon dioxide in the atmosphere caused by the burning of fossil fuels and land-use change. In summary, we introduced our graduate students to the multidisciplinary field of climate system science (see caption for Photo 17.1).

In terms of research, the CRG members worked mainly on the physical and dynamic aspects of Earth's climate system, which was the focus of the World Climate Research Program (WCRP), initiated in 1979 by the World Meteorological Organization. In 1986, in response to growing concerns about both the physical *and* biological state of the global environment, the International Council of Scientific Unions (ICSU) created the International Geosphere-Biosphere Program: A Study of Global Change (IGBP). Shortly after, WCRP and IGBP signed an agreement in a focused quest to understand the linkages between the geosphere and biosphere (a field now called Earth System Science). It became clear that to participate fully in the various projects sponsored by WCRP and IGBP, the CRG needed to expand its membership to include faculty members from other departments, especially those that dealt with the biosphere and biogeochemical cycles (such as the carbon cycle). I was encouraged to move in this direction by Sherry Olson, the then Chair of the Geography Department.

"Lawrence," she said over the phone one day, "I want to arrange a meeting with you and four land-surface colleagues. I think they should collaborate with your CRG members in the areas not represented by your department."

And how right she was! After listening to what research Tim Moore, Ian McKendry, Wayne Pollard, and Peter Schuepp were doing on peat bogs, urban meteorology, high-latitude permafrost, and atmosphere-land gas exchanges, it was obvious that these colleagues would be valuable additions to an expanded CRG.

Thus, in February 1990, I proposed to the McGill senior administration the establishment of a Centre for Climate and Global Change Research (C^2GCR) which would bring together 13 faculty and 38 graduate students from four departments (Meteorology, Geography, Renewable Resources, and Economics) representing the faculties of Science, Agriculture, and Arts [2]. With the specific programs and projects in WCRP and IGPB in mind, the proposed members of C^2GCR (the Centre hereafter) would fall into one of the following four subgroups:

- Global climate modelling and data analysis.
- Biogeochemical and hydrological cycles.
- Small-scale and land-surface processes and their parameterizations.
- Impacts of climate and global change.

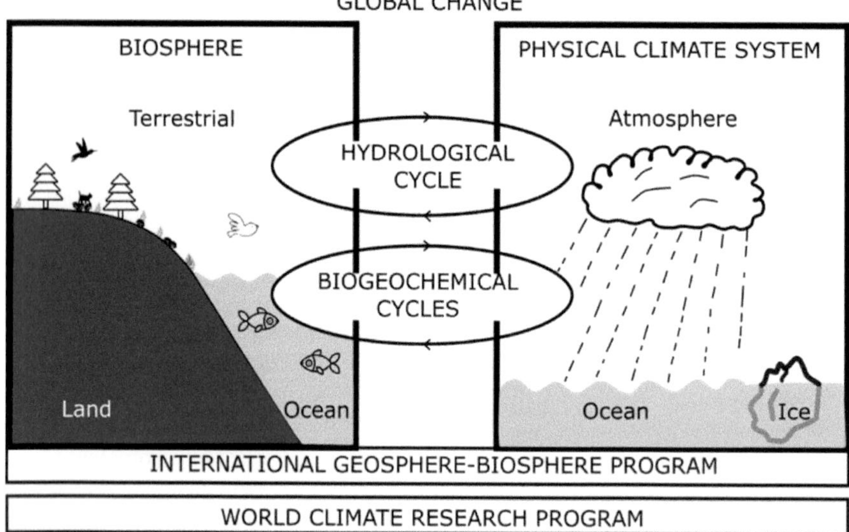

Fig. 17.1 Linkages among physical, chemical, and biological processes, such as the hydrological cycle and biogeochemical cycles, which are critical to our understanding of global change on decade-to-century timescales (figure drawn by Djordje Romanic)

Also, in the manner indicated in Fig. 17.1, it was anticipated that members of the subgroups would become linked together in future interdisciplinary research projects under the global change umbrella. During the last three decades, this has indeed happened. In the case of my own research (and that of my students), I ended up working in all four areas of these subgroups. In addition, I became involved in paleoclimate dynamics research.

Another reason for creating a Centre that would be officially recognized by McGill's Faculty of Graduate Studies and Research is that we would be eligible to apply to the Québec Government's Fonds pour la formation de chercheurs et al.'aide á la recherché (FCAR) for a major grant under the "Research Centres Program." Such a grant would provide major support for graduate and postdoctoral student stipends and research equipment.

Happily, the proposal for C^{2}GCR was approved by McGill's Senate on February 22, 1990, and shortly after, I was formally appointed as founding director of the Centre. Once this happened, the Faculty of Graduate Studies and Research awarded the Centre infrastructure and publication grants that enabled the Centre to expand its activities, develop collaborative research projects and increase its membership. By 1992, there were 15 faculty members in the Centre, which included two atmospheric scientists from Université du Québec à Montréal. The two main objectives of the Centre were (1) to promote research on the interactive physical, biological, chemical, and socio-economic processes that regulate our global environment, and (2) to provide a stimulating academic environment for graduate and postdoctoral students in the emerging fields of earth system science, and climate and global change impacts.

The first objective has its roots in the pioneering work of James Lovelock and his collaborators Dian Hitchcock and Lynn Margulis who regarded the Earth as a living system made up of interdependent relationships [3].

The creation of the Centre also percolated down to the undergraduate level by encouraging instructors to include more climate and global change topics in their courses. Chris Green, a Centre member from the Department of Economics, taught an important undergraduate course called "Economics of Climate Change." The Centre also indirectly led to the introduction of a new Department of Geography course entitled "Global Change: Past, Present and Future." This course was taught by Professors Gail Chmura and Tim Moore, and it attracted around 200 students every year. Much later, around 2005, the Faculty of Science introduced an undergraduate program in "Earth System Science," which proved to be very popular for students with a multidisciplinary interest in environmental modelling and data analysis.

The Centre continued publishing an expanded Technical Report Series (student theses, maps, data reports, literature reviews, preliminary research results, and preprints) and the quarterly newsletter, the C^2GCR Quarterly, which went out to over 400 subscribers. We printed a brochure describing the Centre, which further helped to advertise our activities (Fig. 17.2). The Centre expanded its fortnightly Colloquium Series to include many international speakers. This helped enormously to increase the visibility of the Centre and to bring various members of the Centre together. We also introduced annual "C^2GCR Student Days" and "C^2GCR Faculty Days" at which the Centre students and faculty presented summaries of their recent research accomplishments. As director of the Centre, I submitted an annual report to the Principal of McGill, who was David Johnston at the time. After each report, I was always touched by his handwritten note to me in which he congratulated us on the accomplishments of the Centre. I learned that he was keenly interested in the impacts of global environmental change and chaired the Prime Minister's "National Round Table on the Environment and the Economy." David later went on to be Canada's 28th Governor General.

A major activity of the Centre in 1993 was the preparation of an application to FCAR for a Centre grant, which we knew had to provide evidence of research collaboration among the Centre members. The application was successful, and the funding started in the summer of 1994. This enabled us to hire six postdoctoral students and provide support for graduate student stipends and travel to conferences.

During the late 1980s and early 1990s, many outstanding graduate students and postdocs, as well as sabbatical visitors, were attracted to the Centre. Also, around this time, the Intergovernmental Panel on Climate Change (IPCC) was established by the World Meteorological Organization and the United Nations Environment Program. Thus, the creation of the Centre was very timely, as many of the Centre members later became contributing authors to the Assessment Reports on the Science of Climate Change, which appeared approximately every five years, starting in 1990. I was particularly delighted when I received in 2014 a signed copy of the Fifth Assessment Report co-edited by my former Swiss postdoc Thomas Stocker [4] (see also Chap. 24).

Fig. 17.2 Cover page of the Centre's bilingual brochure, created first in 1992

The success of the Centre over the years can be attributed to the work and support of many individuals and departments. Professor Chris Green (Economics Department) was an outstanding editor of the Centre Quarterly; Chris was ably assisted by Ann Cossette, who also served as the dedicated secretary, receptionist, host, organizer, and student coordinator for the Centre. Professor Nigel Roulet put together an excellent Colloquium Series for the Centre. We had many outstanding speakers from Europe (e.g., André Berger MAE, Victor Brovkin, Martin Claussen MAE) and the USA (e.g., Robert Cess, John Walsh, Carl Wunsch NAS). The Departments of Meteorology (renamed Atmospheric and Oceanic Sciences (AOS) in 1992) and Geography generously provided office and seminar room space for the Centre's operations. The Centre also benefited from the fact that two department chairs, Jacques Derome from AOS and Tim Moore from Geography, were active climate and global change researchers. The operations and activities of the Centre were overseen by a seven-person executive, chaired by the director. This executive included two student representatives who made many helpful suggestions on how to improve the Centre's activities.

As director of C^2GCR I was often invited to give talks about climate change and its impacts on society. Over four decades, right up to COVID-19, I gave nearly 100 public lectures to schools, colleges, universities, national academies, Rotary clubs, senior residences, libraries, government agencies, private clubs and other social organizations. Earlier titles for my talks include "Should we (still) be concerned about climate change?" (thanks to Chris Green for this) and "Climate Change: Where on Earth are we going?" Later I incorporated some of my own research results into the talks, with titles like "Hockey sticks, outdoor rinks and climate change" and "Climate change: From here to eternity." The last title reflected my developing interest in paleoclimates.

Generally, I was well received and appreciated by the audiences for these public lectures, but on occasion I did have to push back against climate skeptics, who often did not accept the fact that most of the warming of the planet during the past century and a half has been caused by the burning of fossil fuels and land-use change. Nevertheless, I did feel that overall I was doing the public a much-needed service with my scientific outreach.

In addition to these public lectures, I was frequently interviewed by the press and on radio and television about recent climate events such as ice storms, floods, forest fires, melting hockey rinks and heat waves. I also battled climate change skeptics on TV; sometimes I think I won the argument about the reality of humankind-induced climate change and at other times, it was a draw. These types of public outreach have also been carried out with much success by former members of C^2GCR. I include in this list Andrew Weaver, Thomas Stocker, Gavin Schmidt, Bruno Tremblay, and Nigel Roulet.

After six years as director of the Centre, I needed a break, and I was delighted that Department of Geography Professor Nigel Roulet, an expert in biogeochemical cycles and climate change, agreed to take over the directorship. In 2000, he organized an excellent symposium to celebrate the 10th anniversary of the Centre, which featured Principal Bernard Shapiro as an invited speaker. We were all highly amused when he said, "This is the first time I've come across an organization whose acronym (C^2GCR) is algebraic."

The directors that followed Nigel include Gail Chmura and Wayne Pollard from Geography, and Charles Lin from AOS. After about 15 years of operation, the Centre changed its name to "Global Environmental and Climate Change Centre" (GEC^3), in response to the Québec Government's suggestion that we broaden the mandate of the Centre to include other Montréal Universities and focus on climate change impacts and mitigation. There was then less support for graduate students. Also, future Québec funding was to be directed towards hiring research professionals and providing infrastructure support. At about this time, OURANOS (a consortium for regional climate change and adaptation) was formed in Montréal by the Québec government to collect/collate data and serve entities interested in climate change. These developments together with the challenge of finding a unifying research theme for all the Centre members proved to be very challenging. Thus in 2015, Nigel Roulet, the last director, and Martin Grant, the then Dean of Science, decided that we should disband GEC^3. While some collaborative research was continued by a few members of the Centre, no new major interdisciplinary climate projects were started.

17.1 Epilogue

While I felt sad when the Centre was disbanded after 25 years of operation, I recognized that I was fortunate to be able to learn so much about different aspects of climate and global change from my colleagues. As mentioned earlier, my own research took off in many different directions over the past three decades. Also, I am very proud of the many students and postdocs that I have supervised at McGill who have now taken on leadership positions in the academic and research world (see Chap. 24). For example, I supervised or co-supervised students who now hold professorial positions in Victoria (Andrew Weaver), New York (David Holland), Montreal (Bruno Tremblay), Zhuhai (China) (Zhaomin Wang), Boulder (Colorado) (Alexandra Jahn), Winnipeg (Juliana Marson), Bern (Thomas Stocker), Hong Kong (Jianping Gan), Brighton (UK) (Yi Wang), and Louvain (Belgium) (Michel Crucifix). One former postdoc, Gavin Schmidt, is the current director of NASA's Goddard Institute for Space Studies (New York). He works tirelessly on the public outreach of science, explaining the current climate issues on many media outlets.

I have recently read Peter Frankopan's treatise entitled "The Earth Transformed: An Untold History" [5], which shows how throughout the history of humankind, environmental and climate change have always played an integral part in the success or failure of past settlements, civilizations, and empires. However, even today these lessons of the past are not taken seriously enough by many of our political and industrial leaders. "I think it's fair to say that I'm fairly pessimistic about where we are right now – but things could change quickly," Frankopan recently said in an interview with Jamie Portman (Montreal Gazette, April 15, 2023). Frankopan is encouraged by the commitment of a new generation of young people determined to make real change in the future. And I am too.

Acknowledgements I thank Janet Boeckh for her comments on the first draft of this chapter and Olivia Marino for assistance with inserting the figures and photograph. I am indebted to Josef Schmidt for his insightful literary comments, and I am grateful to Tim Moore and Nigel Roulet for their input on some of the historical developments of the Centre. I also thank Gail Chmura, Chri Green, Charles Lin, Wayne Pollard, and Richard Thomson for their feedback.

References

1. Mysak LA, Groot C, Hamilton K (1986) A study of climate and fisheries: interannual variability in the northeast Pacific Ocean and its influence on homing migration of Sockeye Salmon. Climatological Bull 20:26–35
2. Mysak LA (1990) Proposal to establish a research centre at McGill University entitled: Centre for climate and global change research, C^2GCR. Department of Meteorology, McGill University. Unpublished manuscript, pp 9
3. Watts J (2015) The many lives of James Lovelock: science, secrets, spycraft and gaia yheory. Greystone Books, Vancouver
4. IPCC (2013) Climate change 2013: The physical science basis. In: Stocker TF, Qin D et al (eds) Contributions of the working group I to the fifth assessment report of the intergovernmental panel on climate change. Cambridge Univ Press, Cambridge, UK and New York, NY, USA, pp 1535 (First published in 2014)
5. Frankopan P (2023) The earth transformed: an untold history. Alfred A Knopf Publisher, New York, NY

Chapter 18
Caught in a Rip Current, Waikiki Beach, Honolulu

Abstract In February 1991, I was participating in a climate workshop in Honolulu. After a long day of intense discussions on the mechanisms for decadal-scale natural climate variability, I decided to take a relaxing swim out to a reef near the New Otani Kaimana Beach Hotel, Waikiki Beach. Getting out to the reef was easy, but for some crazy reason, I started swimming past the reef out to the open ocean, going with the flow of a strong rip current. I suddenly realized that I could not swim against it to return to shore. However, by thrashing my way over to the reef and then slowly clawing over it toward the shore, I managed to survive. I was a mess with much blood on my hands and very lucky there were no sharks nearby!

The New Otani Kaimana Beach Hotel is at the eastern end of Waikiki Beach. The landmark Diamond Head volcano rises steeply just behind the hotel. The setting of the crimson sun is a marvel to watch from the hotel's Hau Tree Lanai Restaurant. Janet (my future wife) and I had a romantic dinner there in February 2014. But I didn't tell her that I nearly drowned in front of the hotel in the early 1990s.

In February 1991, an international climate committee invited a group of us to Honolulu to write a scientific report on our understanding of the mechanisms that produced mid-latitude climate variability on the decadal timescale. The committee asked us to answer the following questions: What is the source of this variability? Is it due to strong El Nino-type oscillations in the tropical Pacific which propagate to the mid-latitudes? Or are strong air-ice-ocean interactions that occur in the Arctic able to force decadal climate changes at lower latitudes?

I and a few others believed in the second mechanism, but a distinguished climate scientist from Boulder, Colorado said, "Mysak, you are full of beans. There is no way those Arctic interactions in such a small area can significantly affect lower latitude climate."

Most of the workshop attendees sided with the Boulder scientist and hence supported the first mechanism. Consequently, I left that afternoon with disappointment written all over my face. I lost the debate. I then retreated to the Kaimana Beach hotel for a relaxing swim.

L. Mysak, *Adventures in Climate Science, Ocean Waves, and the Flute*, Springer Biographies, https://doi.org/10.1007/978-3-032-19848-8_18

In the 1980s, I collaborated with Hawaiian Professor Lorenz Magaard on subsurface Rossby waves in the North Pacific. These waves impact on submarine detection, a topic of obvious interest to the US Navy. Thanks to the Navy's support, my late wife Mary and I were frequent visitors to Honolulu and guests at the New Otani Kaimana Beach Hotel. Lorenz often joined us for dinner at the Hau Tree Lanai Restaurant.

One day, Lorenz said, "Why don't we swim out to the coral reef? It's just 200 m offshore from the hotel."

We did this with ease and rested our bums gently on the edge of the sharp reef. There the waves splashed warm, salty water over our faces. The trade winds cooled our faces. The murmur of the breaking waves was soothing. "What bliss!" I shouted.

In February 1991, my swim to the reef turned into a nightmare. As in the 1980s, I swam out to the reef and rested there. Then I decided it would be fun to go around the corner of the reef and head out to the open ocean. I could see wind surfers having a ball out there, and I wanted to join them. This was an easy swim as there was a rip current that flowed past the side of the reef. I simply swam with the flow, happily washing away my disappointment of the day. It did not occur to me that I might have trouble getting back to shore.

A good swimmer can normally do one body length per second—for a tall person that means 2 m/s. I can keep up this speed for maybe 50 to 100 m. However, rip currents are much stronger, generally being around 4 to 5 m/s. This was certainly the case here. I quickly decided I had better start swimming back to shore. But I could not make any progress. Despite my strongest efforts, I was slowly drifting farther out to sea. I started to panic.

"Is this it for me?" I asked myself. There were no lifeguards or other swimmers near the shore, and so a cry for help would be futile. *I was scared as hell.*

I thought of three possibilities. I might drown right there—I could only tread water for so long. Then I thought I could try swimming along the offshore side of the reef, going parallel to the shore, which would take me toward downtown Honolulu. But once past the reef I would still need to swim back to shore a great distance—about one kilometer. I did not have the strength to do such a long swim. Finally, in desperation, I decided to work my way along the side of the reef toward shore, using my hands to claw myself forward. While doing this, however, I drew blood on my hands, which reddened the water. I looked a mess. I looked over my shoulder to see if a shark was sneaking up behind me.

Thankfully, after about 15 min of this hard and painful work, I did get back to the inshore side of the reef. There I rested my weary body and washed my wounds. My hands stung. The minutes of rest seemed like hours. But I finally swam back to shore from the reef.

To this day, I have never told anyone of this near-death experience. I was foolish to do what I did, especially having studied ocean and wave dynamics. But I learned my lesson.

18.1 Epilogue

This chapter is based on an assignment which I was given in the Thomas More Institute (TMI) Memoir Writing Course that I took in winter 2018. I am very grateful to Pauline Beauchamp and Karen Nesbitt, the course instructors, for their feedback and suggestions for the improvement of this short memoir.

Chapter 19
Presiding Over the Academy of Science: Fostering Intellectual Companionship

Abstract In 1993 I was elected to a three-year term as president of the Academy of Science, the largest of the three academies comprising the Royal Society of Canada (RSC), the country's national academy of the arts and sciences. During these three years I traveled across Canada visiting many cities where there were relatively large numbers of Fellows of our academy. My goals were to meet as many Fellows of the RSC as possible and to encourage them to get involved in the many activities of the Academy of Science and the RSC. These activities included a new Science Policy Committee which I initiated, lecture exchanges with other national academies, scientific conferences associatedwith the annual RSC general meetings, RSC publications, and serving on academy selection committees. There were two noteworthy initiatives during my term of office: an exchange lecture program with the National Academy of Sciences of Ukraine and the Partnership Group in Science and Engineering (PAGSE). The former evolved into a visitors' program for young Ukrainian scientists and scholars to come to Canada for three-month stays in the labs of RSC Fellows; unfortunately, this program had to be suspended when Ukraine was invaded by Russia in February 2022. The second activity brought together the leadership of over 20 Canadian science and engineering societies to promote and advance these fields. PAGSE is still in operation today and is often considered one of the more appreciated initiatives of the Academy of Science.

19.1 Introduction

In spring of 1991, I got a surprise phone call from Ernie McCulloch, past president of the Academy of Science of the Royal Society of Canada (RSC). "Lawrence," he asked, "would you be willing to stand for election as one of the candidates for vice president of the Academy of Science?".

I paused, wondering if this was for real, as I had only been elected a Fellow of the Royal Society of Canada (FRSC) a few years back. Also, I was still recovering from the shock of my then wife Mary having had two operations for breast cancer during the past few months. Fortunately, she had now recovered. Then I remembered that

L. Mysak, *Adventures in Climate Science, Ocean Waves, and the Flute*, Springer Biographies, https://doi.org/10.1007/978-3-032-19848-8_19

when I was elected FRSC in 1986, I promised to get involved in the leadership of the science academy after I moved to McGill. Thus, it was time to step up to the plate. At the same time, I was also somewhat reluctant to take on any new academic activities outside of the university because in the spring of 1990, I became the founding director of McGill's Centre for Climate and Global Change Research (Chap. 17). Hence the pause before answering McCulloch's question. My head was spinning with these many conflicting thoughts.

"Yes," I finally replied, "I am willing to run for vice president." I could hear McCulloch's sigh of relief when he heard my positive response. With the recent reorganization of the Academy of Science's governing structure, there was going to be, for the first time, an election for this position. McCulloch was finding it hard to find Fellows to run in such an election.

19.2 My Term as Vice President, 1991–93

I was duly elected vice president two months after Ernie's call. In September 1991, I joined the 21-member Academy of Science Council, which was chaired by President Chris Barnes. This council organized the academy's various activities. As one of two vice presidents, I was also a member of the Academy of Science Executive, which meant that I worked closely with the executives of the other two academies of the RSC (often called the Society for short). These academies were L' Académie des lettres et des sciences humaines (also known as (aka) Académie I) and The Academy of Humanities and Social Sciences (aka Academy II). The three academy executives, together with the RSC's President, Secretary, Treasurer, Foreign Secretary, and a few others, formed the overarching Council of the Society. I soon discovered that each academy had its own culture and way of doing things, and that distinguished academics liked to have their egos stroked. Consequently, much diplomacy and a tolerance of other viewpoints would be needed to make the Society work for the benefit of all three academies.

Barnes was an inspiring and enthusiastic president of the Academy of Science/L' Académie des sciences (aka Academy III/Académie III). His mission was to elect the most distinguished Canadian scientists to the academy and then entice them to be involved in its activities and projects. In the 1990s these included, for example, the Canadian Global Change Program, exchange lectureships with other national science academies, the publication of overviews on cutting-edge science, and the promotion of public awareness of science. Also, Barnes was very keen on having the academy be recognized by the Federal Government and the public at large as Canada's National Academy of Science, a source of science expertise and advice. For many years, the Science Fellows had wanted to play this role [1].

Shortly after Ukraine gained its independence on August 24, 1991 and because of my Ukrainian ancestry, I had a surprise visit from Academician A G Sitenko, Director of the Institute of Theoretical Physics of National Academy of Sciences of Ukraine (NASU). Dr. Will Zuzak, a physicist friend also of Ukrainian descent, brought Sitenko to my McGill office to discuss possible exchange agreements between the RSC's Academy of Science and NASU. Sitenko pulled out of his briefcase several folders for me to look over. He thought that as vice president of the Canadian science academy, I had the authority to immediately sign these proposed exchange agreements! "Please sign here," he said. Little did he realize that we were a poor academy, with nothing like the stature and resources of NASU. Suddenly I remembered that the RSC had an exchange lecture agreement with the Royal Society of London, and thus I proposed that we could try to set up a similar exchange with NASU. However, I knew there would be a problem with financing and sustaining such an exchange.

19.3 Election as President of the Academy of Science, Spring 1993

After two years as vice president, I was elected to a three-year term as president of the Academy of Science. I was happy to learn that my opponent in this election, physicist John Hardy, would continue to serve on the academy executive as a vice president for two more years. As a prominent scientist at Atomic Energy of Canada Limited, his expertise in dealing with "big science" issues would be very valuable in the academy council deliberations. In the same election cycle for new members of Council, Michel Chrétien, a distinguished medical science researcher (and brother of the then prime minister, Jean Chrétien), was elected as the second vice president.

At first, I was quite intimidated by my position as president. However, Dr. McCulloch gave me some helpful advice at a private dinner before I officially became president. "When you are struggling to deal with a tough issue," he said, "consult your academy council. There is a lot of wisdom around the table, and the council members will be eager to help you solve the problem at hand."

19.4 Brief History of the Royal Society of Canada and the Academy of Science

The Royal Society of Canada was founded in 1882 by the then Governor General of Canada, The Right Honorable Marquess of Lorne. The founding president was McGill Principal John William Dawson, an eminent paleontologist. The main objective of the Society is the promotion of learning and research in the arts and sciences. At the beginning, its 80 elected Fellows were grouped into four 20-member sections: French literature; English literature; Mathematical, Physical and Chemical Sciences; and Geological and Biological Sciences. Over the years these sections subdivided into more specialized fields, and then in 1974, the many different sections were regrouped into Academies I, II, and III, as defined above [1].

Early initiatives of the Fellows of the Society led to the founding of the Marine Biological Station in St. Andrews, NB, in 1899, and the National Museum of Canada in Ottawa, in 1927. After World War II, there was a strong desire of the Fellows in Academy III to achieve the status and prestige of other national science academies such as the Royal Society of London (RSL) and the US National Academy of Sciences (NAS), which regularly provided timely information and advice to the government and the public. In contrast to these well-funded and elegantly housed academies in London and Washington, DC, respectively, during the 1990s our Academy of Science tagged along with the other RSC academies in a rented suite in an office tower of dentists and doctors in downtown Ottawa. The funding for our Ottawa staff and basic operations (electing new Fellows, organizing the annual meeting, awarding medals and prizes, sponsoring conferences) came from membership dues and a small endowment.

In 1989 the Society was fortunate to have received from the then Ministry of Industry, Science and Technology Canada a five-year infrastructure grant of one million dollars per year to carry out many projects that involved members of Academy III, such as Public Awareness of Science, Promotion of Women in Scholarship, and Evaluation of Research. However, if Academy III wanted to be active in providing scientific advice to government and be better known to the public (like the UK RSL and the US NAS), more outreach activities would have to be initiated by the academy leadership and a greater participation of the Science Fellows would be needed, just as President Barnes had desired. Also, a plan for enhanced funding would have to be developed.

19.5 My Term as President, 1993–96

The above events and ideas prompted me, as the new president and chair of the academy council, to write an open letter (in English and translated into French) to all members of Academy III in July 1993. After hearing how much the academy Fellows appreciated this letter, I followed up with a similar missive at the end of each year of my presidency (1994, 1995, and 1996). In the July 1993 letter, I proposed the following action items:

1. Council members from the different scientific divisions of the academy would prepare short position papers on "hot" scientific topics and bring these to the attention of the relevant government ministers.
2. The academy executive would approach distinguished Canadian and international scientists to speak at the science council meetings, which were held every four months. Scientists to be invited included the presidents of the Natural Sciences and Engineering Research Council and National Research Council, and the Foreign Secretary of the US NAS.
3. Establishment of an Academy III Science Policy Committee that would include members of the other academies.

4. Preparation of a bilingual, five-panel pamphlet describing Academy III and its place in the Royal Society of Canada (Fig. 19.1).
5. Visits (by myself) to cities across Canada where there are substantial numbers of Academy III Fellows. This would allow me to personally meet and exchange views and concerns with Fellows outside the academy council.
6. Development of lectureship exchanges with other national academies, as an expansion of the exchanges we had with the Royal Society of London and the Institut de France.

Regarding item 3 above, I was happy that Keith Brimacombe, a distinguished metallurgical engineer from UBC, agreed to chair the Science Policy Committee. In discussing possible members for the committee, the name of William Leiss, then at Simon Fraser University, came up, as he had done consulting work on the impact of

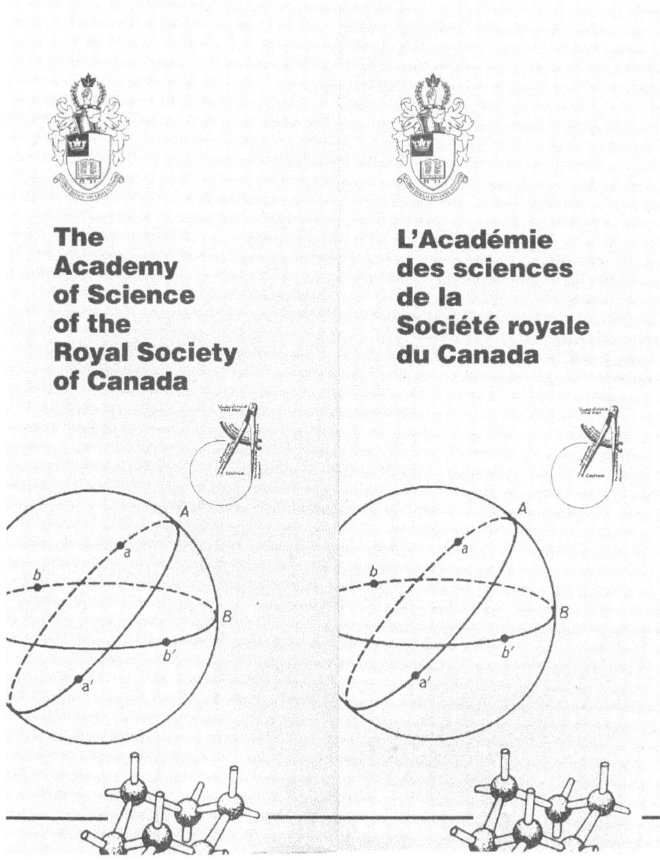

Fig. 19.1 The cover panels (English and French) of the bilingual Academy of Science pamphlet, produced in December 1993 by Janice Klein, the Academy of Science Coordinator. Permission granted by the Royal Society of Canada

environmental pollution on health. At the 1993 Annual General Meeting (AGM) of the RSC in Montreal, I had an informal chat with Leiss, whom I had met earlier at a UBC conference, about joining our Academy III-led Science Policy Committee. He was delighted to do so.

This was my first blunder in working with the other academies of the Society because Leiss was a Fellow of Academy II. A day later at the AGM, Craig Brown, the outdoing president of Academy II, handed me a hand-written letter. "Professor Mysak," he wrote, "it is only the officers of Academy II who appoint their members to inter-academy committees. Please remember this in the future."

I was upset when I read Brown's terse letter, but to keep the peace, I quickly wrote him a note apologizing for my inappropriate action. Fortunately, he accepted my apology, and Leiss was later formally approved by the Academy II executive for service on the Academy III-led Science Policy Committee. This was lesson 1 in inter-academy diplomacy.

One important action by the Committee chaired by Brimacombe was the preparation of an invited response to the "Discussion Paper" put out in 1993 by Industry Canada as part of its Science and Technology Review of federal spending on science. In our response, it was recommended that the federal science departments, such as Environment Canada, Fisheries and Oceans, and Natural Resources Canada, seek the advice of Canada's distinguished scientists about where and how these departments could make the best use of their money to support basic and applied research. As a move in this direction, these and other science departments did set up advisory panels chaired by deputy ministers to do just that in the late 1990s. I know many Fellows of Academy III happily served on these panels, and I was invited to be on the Environment Canada advisory panel for the three-year period 2000–03. At the beginning, I personally felt that our advice to the Deputy Minister of the Environment was much appreciated. However, by the end of my term, it seemed that the government was becoming increasingly disinterested in what we had to say, and indeed, all these advisory panels were disbanded a few years later.

I really enjoyed and benefited from my visits across Canada where in each locale, a luncheon or dinner was held with the Fellows of Academy III. I was usually featured as a guest speaker and had a lively Q&A session after my talk. In some of the smaller cities, Academy I and II Fellows joined us in these get togethers. In addition, newly elected Fellows were often invited to give public lectures about their work; these sessions were very popular and helped to foster the intellectual companionship that many Fellows were seeking. From my visits across Canada, I learned that most scientists are quite sociable and far from being loners. During my term as president, I made over such 20 visits, from coast to coast. In preparation for each visit, I worked with a local Fellow (preferably a dean or vice president) to make the arrangements and help pay for the dinner, advertising, accommodation, etc. From the science academy's modest budget, I used funds to pay for wine at these dinners. It is a pleasure to thank again all those who helped make these arrangements. Sadly, many of these Fellows are no longer with us.

Reflecting on these visits three decades later, I think many Fellows would appreciate having Canada-wide "local chapters" of the Society which could organize events and meetings a few times each year. I know from personal experience that I have benefited from the activities sponsored by the local chapters of Canadian Meteorological and Oceanographic Society of Canada, especially when I could not attend its AGM. However, for an RSC local chapter program to work well, regional Fellows must be willing to take on organizational roles, *and* some Society funding and staff support would be needed. The latter two issues are certainly problematic today since the Society does not have sustained infrastructure funding from the Federal Government and the Ottawa office now has shrunk to about seven staff members.

I still remember the shock the Society executive received on December 23, 1993, when a letter was delivered by hand from the Ministry of Industry, Science and Technology, informing us that the five-year infrastructure grant would not be renewed in 1994. Further, it was announced that we would not be receiving the last quarterly installment of $250,000 from this grant. Later we learned, however, that the Society would continue to receive 100 K per year for the Public Awareness of Science program. Nevertheless, the ending of the infrastructure grant from this ministry meant that large budget cuts were necessary and that the RSC staff of over 20 would have to be substantially reduced. It was a sad day in the office. The Academy of Science budget was slashed from 100 to 25 K per year. Fortunately, I still had the support of the Academy of Science Coordinator Janice Klein for another two years. She ably helped the Academy with many of its projects, including the production of its brochure (see Fig. 19.1). At the end of my second year as president, Janice was transferred to other duties. Then the academy was lucky to have the assistance of Nancy Schenk, who at the same time served as the Executive Secretary of the Society President (Robert Haynes, for the term 1995–97).

Chairing the academy council meetings was generally an enjoyable task, and most issues were readily resolved by open and frank discussions. However, during my second year, I ran into a tricky situation over the election of new Fellows to the academy. In the 1989 reorganization of the academy Fellow selection procedures, each academy divisional selection committee—there were four of them—was to provide Council an A list and a B list, with each list having the number of candidates equal to the quota set for that division. The A-listed candidates formed a top-tier group, and the B-listed candidates were a second-ranked group, who all, however, deserved the FRSC. These lists from the four divisions were next reviewed by an academy-wide selection committee which then chose a mixture of A- and B-listed candidates to make up the academy quota of new Fellows for the year (24 at the time). In principle, this committee tried to get a wide representation of new Fellows from across the country. This procedure really upset the divisional directors because they felt some very highly ranked A-listed candidates were left off the ballot that was later submitted to the whole academy for a vote. I had a mutiny on my hands!

Fortunately, our able academy secretary, Michael Dence, suggested a Solomonic resolution to this problem. For each division, a fixed number of A-listed candidates would go on the final ballot, with the total number adding up to 18. Then a smaller number of B-listed candidates (12) would go into a pool which would be reviewed by the academy-wide selection committee. From this pool of candidates, six would be chosen to go on the final ballot, which would contain 24 names, the quota for election to the academy. The six additional candidates were chosen using the guidelines of gender, representation from small universities, and unrepresented fields. I was very happy that the Council accepted this very Canadian resolution. As far as I know, the academy still uses this pool system to make up the final list of candidates to be voted on by all the academy members. However, today the final list of candidates has about 50 names, and the Science Fellows only vote for new members in their division.

During my term, I worked with two Society presidents, John Meisel (1992–95), a political scientist from Queen's University, and Robert Haynes (1995–97), a geneticist from York University. Meisel wanted the Society's projects to be quite broad and be the most visible part of our organization. Perhaps for political reasons, he was not comfortable with some of the initiatives of Academy III. Haynes, on the other hand, wanted each of the academies to have more visibility and to develop distinctive activities. For the 1996–97 Calendar, which listed all the Society and academy committees as well as the members, he requested that the names of the academies be listed on the front cover. The RSC staff members disliked this change, and they made sure that such a cover never appeared again.

One of my more rewarding projects with the academy was working with Ukrainian Canadian physicist Jurij Darewych to raise funds (an endowment) for a lecture exchange with the National Academy of Sciences of Ukraine (NASU). Under Darewych's guidance, we were able to secure a major grant from the Ukrainian Canadian Foundation of Taras Shevchenko. In addition, many Fellows of the RSC and individuals and associations from the Canadian Ukrainian diaspora donated generously to support the exchange. We soon had an endowment for this lecture exchange of over $40,000, and the interest from this was enough to support in 1995 the first lecturer from Kyiv, Professor Anatoly Zagorodny. He was Deputy Director of NASU's Bogolyubov Institute of Theoretical Physics and internationally known for his work on plasma physics. With the organizational help of Josée Lalande at the RSC office, Jurij Darewych and many local Fellows, Dr. Zagorodny was able to visit Toronto, Montreal, Ottawa, Winnipeg, Saskatoon, Edmonton, Vancouver, and Victoria, where he gave lectures to physics departments and addressed various Ukrainian professional and business organizations about science and education in Ukraine. We were thrilled that this inaugural lecture exchange was so successful.

On the international front, I also had much pleasure in making courtesy visits to other national science academies, where I was often hosted by the president or foreign secretary. Usually these visits occurred when I was attending a conference. Among the academies visited were the Royal Society (London), the National Academy of Sciences (USA), the Belgium Royal Academy of Sciences, the Czech Academy of Sciences, The Swedish Royal Academy of Science, and NASU.

Another exciting initiative of the Academy of Science in June 1995 was the formation of the Partnership Group for Science and Engineering (PAGSE), a gathering of the presidents of over 20 Canadian science and engineering societies together with three academy executive members (Michael Dence, Howard Alper and me). During 1995–96, PAGSE discussed various issues of common interest, drafted an action plan, and prepared a response to the federal government's Science and Technology

Strategy. Also, as part of the 1996 AGM of the Society, PAGSE hosted a public conference entitled: "Science Highlights/La Science en Vedette," with speakers sponsored by the member organizations of PAGSE (Fig. 19.2). We were delighted that over 150 people attended this conference.

During my three years as president, there were many other responsibilities that Academy Council had to contend with, such as seeking Fellows to write the obituaries of the deceased Fellows, persuading academy Fellows to serve on various

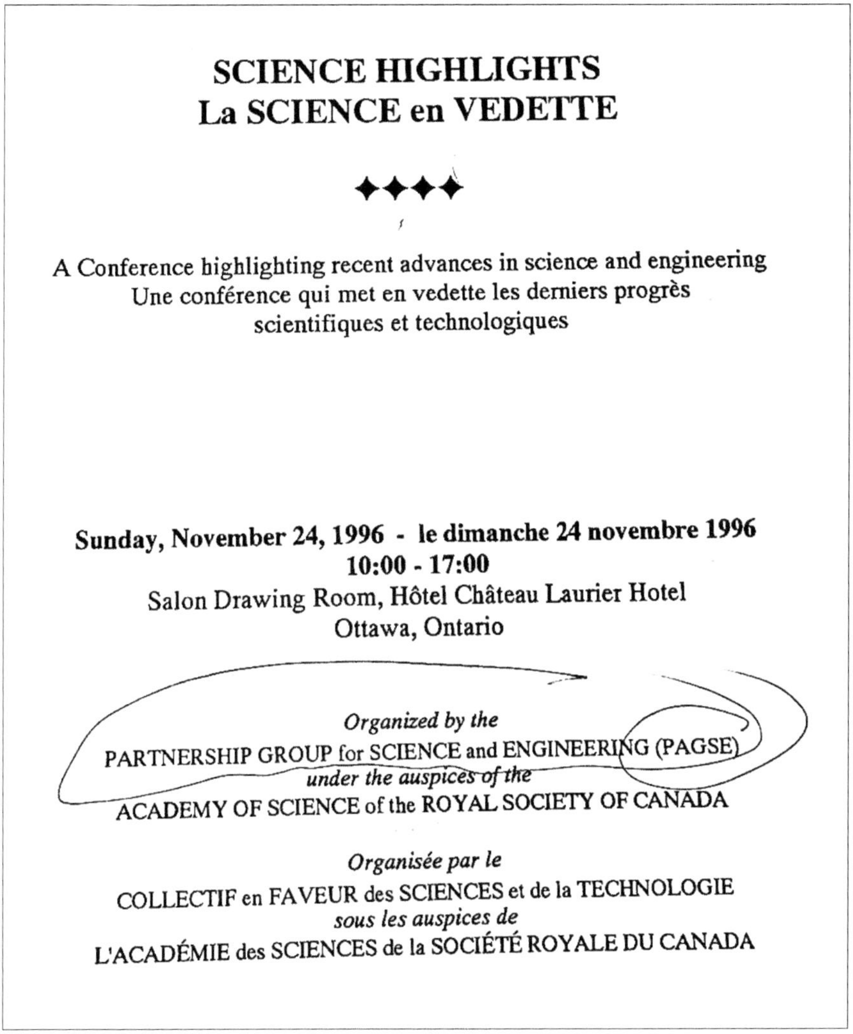

SCIENCE HIGHLIGHTS
La SCIENCE en VEDETTE

A Conference highlighting recent advances in science and engineering
Une conférence qui met en vedette les derniers progrès scientifiques et technologiques

Sunday, November 24, 1996 - le dimanche 24 novembre 1996
10:00 - 17:00
Salon Drawing Room, Hôtel Château Laurier Hotel
Ottawa, Ontario

Organized by the
PARTNERSHIP GROUP for SCIENCE and ENGINEERING (PAGSE)
under the auspices of the
ACADEMY OF SCIENCE of the ROYAL SOCIETY OF CANADA

Organisée par le
COLLECTIF en FAVEUR des SCIENCES et de la TECHNOLOGIE
sous les auspices de
L'ACADÉMIE des SCIENCES de la SOCIÉTÉ ROYALE DU CANADA

Fig. 19.2 Program cover of the conference sponsored by PAGSE, which highlighted recent advances in science and engineering made by Canadian researchers. Permission granted by the Royal Society of Canada

committees, and encouraging Fellows to nominate their outstanding colleagues for FRSC. Unfortunately, too many newly elected Fellows just loved to have the FRSC initials after their name and didn't want to serve on any Society committees. I found serving as president an enriching experience, and I was touched when President Haynes presented me with a certificate of appreciation at the end of the 1996 AGM (Photo 19.1).

Photo 19.1 Society President Robert Haynes (right) presenting me with a certificate of appreciation upon the completion of my term as science academy president at the annual RSC banquet, November 22, 1996. Photo taken by my then wife Mary Mysak

19.6 My Term as Past President, 1996–99

During these years, my main job was to mentor the new academy president, Geoff Flynn, a medical science researcher from Queen's University. Geoff always had an upbeat disposition, and he welcomed my advice in dealing with challenging issues. I also served as chair of the academy-wide selection committee which had the responsibility of selecting the required number of names from the B-list to be on the final ballot. This was never an easy task, but I was happy when we did manage to get an increased number of women on the final ballot.

I also wanted to ensure that the lecture exchange with NASU, which began in 1995, went well. During 1996, I was delighted that Professor André Bandrauk, a chemist from University of Sherbrooke, was the first Canadian chosen to visit Ukraine and lecture at several chemistry and physics institutes of NASU. During the next year, Professor Platon Kostyuk, a biochemist from Kyiv, came to Canada and gave lectures in Montreal, Toronto, and Ottawa. This lecture exchange of distinguished professors continued until 2007. After this time, the program shifted gears by inviting younger scientists from Ukraine to spend three months in the lab of a senior Canadian scientist, who was an FRSC. However, since the COVID-19 pandemic (2020–22) and the brutal invasion of Ukraine by Russia in February 2022, this program has been suspended.

19.7 Epilogue

Much has changed in the Society and the academies since my term as president three decades ago, but I very much treasure the friendships and contacts that I made in the 1990s. And I still get much pleasure in nominating younger colleagues for FRSC and seeing some of these new Fellows get involved in the Society activities. Recent changes which have benefited the Society are (1) the reorganization in 2007 of Academies I and II into two bilingual academies, one for the creative arts and humanities, and one for the social sciences, (2) the creation in 2014 of the College of New Scholars, Artists and Scientists, (3) the purchase of Walter House (about a decade ago) as a permanent home of the Society (Photo 19.2), and (4) the agreement between the RSC and Canadian Science Publishers that the open access journal FACETS be the official journal of the Academy of Science.

On the downside, it is sad for me to see that the academies now have fewer meetings of their councils and hardly any visible activities. The long-term science academy goal of providing advice and information to government has been taken over by the Canadian Council of Academies (a consortium of representatives from the RSC, the Canadian Academy of Health Sciences, and the Canadian Academy of Engineering), and more recently, by the Office of the Chief Science Advisor, established in 2017. Without sustained infrastructure funding from the Federal Government, the RSC, and the Academy of Science in particular, would never be able to carry out major scientific evaluations and studies such as done by the RS (London) and NAS (US).

Photo 19.2 Walter House at 282 Somerset Street West in Ottawa, the permanent home of the Royal Society of Canada. This classic Victorian house was purchased with the aid of a major donation by William Leiss, a former president of the Society (1999–01). Photograph courtesy of Royal Society of Canada

But to end on a positive note, it is satisfying for me to learn that during the COVID-19 pandemic, the RSC was very successful in (1) setting up a Task Force to help Canadians become a better society coming out of the pandemic, and (2) establishing a partnership with Canada's national paper, the Globe and Mail, to extend the reach of information gathered by the COVID-19 Task Force [2]. These latter two activities certainly make me proud to be a very senior but still interested Fellow of the Royal Society of Canada.

Acknowledgements During my term as academy president, I was very fortunate to have the considerable assistance of Ann Cossette (at McGill University) and Janice Klein (at the RSC Ottawa office) in dealing with the academy work. Janet Boeckh provided excellent feedback on a first draft of this memoir, and Josef Schmidt, my literary critic, offered many suggestions for improvement. I also thank Olivia Marino for inserting the figures and photographs.

References

1. Levere T (1998) Research and influence: a century of science in the royal society of Canada. Reprinted by the royal society of Canada from Scientia canadensis, March 1998, pp 99
2. Royal Society of Canada Annual Report for 2023. Available from the RSC's Walter House in Ottawa, ON

Chapter 20
A Job Interview: Navigating Strong Academic Currents

Abstract In 1998 I was invited to apply for the position of Vice President for Research and Advancement at the University of Alberta, in Edmonton. Being a graduate of the U of A, I saw this as an excellent opportunity to give back to my alma mater, which gave me an excellent start to my five-decade academic career. I had completed my terms as founding director of the climate centre (C^2GCR) at McGill and president of the Academy of Science of the Royal Society of Canada. These positions provided me with administrative and team-building experience and gave me exposure to many leading scientists and scholars across Canada. I ended up as finalist for the position; the president of the U of A chose the internal candidate instead. Although very disappointed with this result, it did prove to be another turning point in my life. I got into new and exciting challenges in climate modeling and paleoclimate research, and I built up an outstanding group of graduate and postdoctoral students. Over the next decade, my research and that of this group was recognized with many awards and prizes, which would never have happened had I gone to the U of A in 1998.

As an undergraduate in applied mathematics in the late 1950s at the University of Alberta (U of A) in Edmonton (a western prairie city), I had a dream. After graduation in 1961 and the completion of a PhD in mathematics elsewhere, I would return to my alma mater as a young assistant professor and "work my way up the ladder." I would become a full professor by the time I was around 40, chair of the mathematics department five years later, and perhaps then dean of the arts and science faculty. I greatly admired the then dean (Walter Johns) when I started at the U of A in 1957, and even more so when he became president in 1959. Johns served admirably in the latter position for 10 years, when the university registration soared from 5000 to over 17,000 and the school of graduate studies flourished. He was a role model for me.

President Johns and his wife came to all the concerts of the University of Alberta Symphony in which I was both principal flutist and assistant conductor. After one of the concerts, Mrs. Johns said "Lawrence, you have a real talent for conducting, and I strongly encourage you to expand your activity in this field." How cool was this, coming from the president's wife!

L. Mysak, *Adventures in Climate Science, Ocean Waves, and the Flute*, Springer Biographies, https://doi.org/10.1007/978-3-032-19848-8_20

My life, however, did not quite go according to this dream. I did complete a PhD in applied mathematics at Harvard University in 1966 (Chap. 7). Since my thesis research was in the application of mathematics to ocean waves, I was attracted to UBC's renowned Institute of Oceanography (IOUBC) which had a strong physical oceanography group led by Robert (Bob) Stewart OC FRS, whom I met at a conference in Moscow in 1966 (Chap. 8). As there were no positions for mathematicians in the Institute, he approached the mathematics department to see whether there were any openings at the assistant professor level. Fortunately, the department had just been allotted 12 new tenure-track positions, and in due course, with the strong endorsement of Stewart, I was offered and accepted one of these. After my arrival at UBC in September 1967, I was also appointed as an associate member of IOUBC where I could collaborate with other oceanographers in my field and give guest lectures on ocean waves in the institute's graduate courses. In this way I attracted new graduate students to work in the application of mathematics to ocean waves and dynamics.

At UBC I did make my way up the academic ladder to professor of mathematics and oceanography in 1976 at age 36. However, I had a young family to support (my son Paul was born in 1975, and my daughter Claire came along in 1978), and there was no way I could consider taking on any administrative position soon. Besides, at the time I was very pre-occupied with my research and co-authoring with Paul LeBlond the treatise *Waves in the Ocean* (Chap. 13).

My Zürich sabbatical in 1982–83 (Chap. 14) was a pivotal year—I learned about El Niño and the impact of interannual climate variability on Sockeye salmon migration in the northeast Pacific Ocean. A few years later, this interest in climate science led to my move in 1986 to McGill where I took up an NSERC Industrial Chair in Climate Research (Chap. 15). Not knowing anyone in Montreal, the first couple of years were difficult for my then wife Mary, but she later enjoyed her time in Quebec and at McGill, where she graduated in 1999 with an honors degree in eastern religions. I thrived at McGill, starting research in several areas of climate science and supervising excellent graduate students. In 1990 I had my first experience of (low-level) university administration when I became founding director of the Centre for Climate and Global Change Research (C^2GCR) (Chap. 17). The Centre flourished, both nationally and internationally, and we hosted many renowned scientists over the next decade. In the middle of my term as C^2GCR director, I also served a three-year term as president of the Academy of Science of the RSC (Chap. 19), which I found very rewarding and stimulating as I got to know the crème de la crème of the Canadian science community from coast to coast.

Soon my reputation as an excellent scientist and capable team-builder became widely known, and I received, in the 1990s, over half a dozen enquiries from other institutions asking me whether I would consider a move into a higher administrative position. I was tempted to apply for such openings at Western University, the Universities of Victoria, Michigan and Saskatchewan, as well as for high-level administrative positions with Environment Canada and the World Meteorological Organization. However, McGill was treating me well and I was happy to continue with my teaching, research, and supervision of graduate and postdoctoral students.

By chance, in winter 1997–98, I learned that my alma mater, the U of A, would soon have an opening for the position Vice President (VP) for Research and Advancement. This seemed to be an odd combination for such a high-level position at a leading research-oriented university, but nevertheless, several colleagues in mathematics and meteorology at the U of A nominated me for this vice presidency. I submitted my CV, along with the names of three references, and shortly after I was invited for an interview with the advisory committee for this position. I spent a lot of time preparing for this interview thinking about how I could advance the research reputation of the U of A. While I felt I was well qualified for the research side of this position, the advancement side was more of a challenge. But I had success in the past raising funds for an RSC-Ukraine Lecture Exchange with the National Academy of Sciences of Ukraine, and I did believe I could handle and enjoy advancement activities for the university. I've always enjoyed meeting new people in all walks of life.

Rod Fraser, the then U of A president, chaired my two-hour interview with the advisory committee. Later I learned from my U of A colleagues that the interview had gone well—I could answer 95% of the questions and many on the committee thought I would be an excellent choice for the job. My main rival for the position was a former dean of the business faculty. He had had great success in getting the funding for several endowed professorships in his faculty; however, he was not known for any notable research accomplishments. In the end, the president favored the former dean. He was the "inside" candidate, and the president liked his strong record in advancement.

I later learned that the president's appointee to this position only stayed there for a couple of years, perhaps because he felt that he was not suitable for the research side of the job. The president also later told a colleague that maybe it would have been better to have appointed Mysak to this position in the first place.

After being formally turned down for the VP position in a phone call from the president, it was later proposed that I might consider, as a consolation prize, coming to the university as a prestigious University Professor. However, I never received a formal offer for such a post, and I'm not sure that I would have accepted it in any case.

I was crushed by the president's decision and disappointed that my dream of giving back to the U of A as a senior administrator was not going to be realized. It took me a year to get over this disappointment and to find a way to move on. "You are in a very grouchy mood these days," my late wife Mary used to say. However, from the family viewpoint, it was just as well that I did not get the post. Claire was beginning her drama studies at Bishop's University (in Lennoxville, in Quebec's Eastern Townships), and Paul was working and studying parttime for a commerce degree at Montreal's Concordia University and still living at home. If I had gone to the U of A, Mary would have stayed behind in Montreal for at least a year to provide a stable environment for the children. It would not have been easy for me to be on my own and start a new job in Edmonton, where I had not lived for over three decades.

The turning point to get me out of my doldrums happened at the 1999 fall meeting of the American Geophysical Union in San Franscico. There I met Martin Claussen MAE from the Potsdam Institute for Climate Impact Research in Germany. He had just given an elegant, invited paper on a new class of climate models (the so-called intermediate complexity models) that lie between the complex 3-D general circulation models of the climate system and the simpler dynamical system or box models that have been around for some time [1, 2]. When I told him about the recent thesis of my PhD student Zhaomin Wang on a simplified earth system model that coupled together the atmosphere, ocean, sea ice, and land surface [3], Claussen immediately invited us to a workshop at the 2000 European Geophysical Society (EGS) meeting in Nice, France. At this workshop, Zhaomin and I were delighted to meet several other groups in Europe who were also developing a similar class of intermediate complexity models. I particularly enjoyed working at this workshop with my longtime colleague and friend André Berger MAE, from Belgium. We all found this workshop very stimulating and agreed to meet at next year's EGS meeting for further discussions. The first outcome of these workshops, a paper by Martin Claussen MAE, me and 17 others which appeared in 2002 [4], has proved to be my most cited paper (over 700 citations). Subsequent papers emanating from these EGS workshops focused on model intercomparison studies involving thermohaline circulation hysteresis [5] and CO_2-doubling experiments [6].

Back at McGill, I was able to attract many new graduate students in earth system science who also wanted to learn more about paleoclimates and the role of vegetation and biogeochemical cycles in the climate system. My former PhD student Zhaomin played a key role as a co-supervisor in much of the research I did with my students during the next decade. We covered many areas of climate and paleoclimate research, publishing papers on glacial inception [7, 8], the cold-climate thermohaline circulation [9], long-term future climate change [10], vegetation dynamics during the Holocene [11, 12], and carbon cycle dynamics in northern Canada [13]. Although I officially retired in 2010, the momentum of my research carried me forward for nearly another decade, with my last PhD student graduating in 2017.

The year 2006 was a rather magical one for accolades recognizing the research accomplishments of my group. At the 2006 European Geosciences Union meeting in Vienna, I was awarded the Alfred Wegener Medal (Fig. 20.1) and Honorary Membership in EGU "in recognition of leadership in oceanography and fundamental contributions in ocean dynamics, sea ice and climate". Later in the fall, my family gathered in Quebec City where I was presented with the prestigious Prix Marie-Victorin for natural sciences by the Government of Quebec. The latter prize is one of the 11 Prix du Quebec awarded annually in all areas of culture and science. And ironically, in 2009, a decade after my interview in Edmonton, I was honored with the Distinguished Alumni Award from the University of Alberta. Clearly, had I been offered the job at the U of A in 1998, I would not have attended any of these memorable award ceremonies. Looking back over two decades later, I can honestly say that today I'm happy for having turned the page the way I did in 1999 by pursuing with vigor a new phase of my career in research and the mentoring of many fine graduate and postdoctoral students.

Fig. 20.1 Left: Alfred Wegener medal awarded to me in Vienna at the 2006 Annual Meeting of the European Geosciences Union. Wegener was a German meteorologist and geophysicist who formulated the first complete statement of the continental drift hypothesis. Right: Inscription on back of medal. The medal came as a complete surprise for me. The then president of EGU asked, "Have you ever received a 'gong' before?" "Yes," I said, "but never one of this size." The medal measures 8 cm across

Acknowledgements I thank Janet Boeckh for her review of a first draft of this chapter and Josef Schmidt for his insightful critique. The comments of Gordon Swaters are also gratefully appreciated.

References

1. McGuffie K, Henderson-Sellers A (1997) A climate modelling primer, 2nd edn. Wiley, New York
2. Stocker T (2011) Introduction to climate modelling. Springer, Heidelberg, Dordrecht, London, New York
3. Wang Z, Mysak LA (2000) A simple coupled atmosphere-ocean-sea ice-land surface model for climate and paleoclimate studies. J Clim 13:1150–1172
4. Claussen M, Mysak LA, Weaver AJ, Crucifix M, Fichefet T, Loutre M-F, Weber SL, Alcamo J, Alexeev VA, Berger A, Calov R, Ganopolski A, Goose H, Lohmann G, Lunkeit F, Mokhov II, Petoukhov V, Stone P, Wang Z (2002) Earth system models of intermediate complexity: closing the gap in the spectrum of climate system models. Clim Dyn 18:579–586
5. Rahmstorf S, Crucifx M, Ganopolski A, Goose H, Kamenkovich I, Knutti R, Lohmann G, Marsh R, Mysak LA, Wang Z, Weaver A (2005) Thermohaline circulation hysteresis: a model intercomparison. Geophys Res Lett 32:L23605. https://doi.org/10.1029/2005GL023655
6. Petoukhov V, Claussen M, Berger A, Crucifix M, Eby M, Eliseev AV, Fichefet T, Ganopolski A, Goosse H, Kamenkovich I, Mokhov I, Montoya M, Mysak LA, Sokolov A, Stone P, Wang Z, Weaver AJ (2005) EMIC Intercomparison Project (EMIP—CO_2): comparative analysis of EMIC simulations of climate, and of equilibrium and transient responses to atmospheric CO_2 doubling. Clim Dyn 25:363–385. https://doi.org/10.1007/s00382-005-0042-3
7. Wang Z, Mysak LA (2002) Simulation of the last glacial inception and rapid ice sheet growth in the McGill Paleoclimate Model. Geophys Res Lett 29(23):2012. https://doi.org/10.1029/2002GL015120

8. Wang Z, Cochelin A-S, Mysak LA, Wang Y (2005) Simulation of the last glacial inception with the green McGill Paleoclimate Model. Geophys Res Lett 32:L12705. https://doi.org/10.1029/2005GL023047
9. Wang Z, Mysak LA McManus JF (2002) Response of the thermohaline circulation to cold climates. Paleoceanography 17(1). https://doi.org/10.1029/2000PA000587
10. Cochelin A-SB, Mysak LA, Wang Z (2006) Simulation of long-term future climate changes with the Green McGill Paleoclimate Model: the next glacial inception. Clim Change 79:381–410. https://doi.org/10.1007/s10584-006-9099-1
11. Wang Y, Mysak LA, Wang Z, Brovkin V (2005) The greening of the McGill Paleoclimate Model. Part I: improved land surface scheme and vegetation dynamics. Clim Dyn 24:469–480. https://doi.org/10.1007/s00382-004-0515-9
12. Wang Y, Mysak LA, Wang Z, Brovkin V (2005) The greening of the McGill Paleoclimate Model. Part II: simulation of natural millennial-scale variability during the holocene. Clim Dyn 24:481–496. https://doi.org/10.1007/s00382-004-0516-8
13. Wang Y, Roulet NT, Frolking S, Mysak LA (2009) The importance of Northern Peatlands in global carbon systems during the Holocene. Clim Past 5:683–693 (Corrigendum 5:721–722)

Chapter 21
My Cutting-Edge Pickup Line

Abstract In January 2012, one month after my wife Mary suddenly died, I was the host of a public "cutting-edge" lecture on brain imaging at McGill. At the reception after the lecture, I introduced myself to Janet and Judy who came to this monthly lecture series for the first time at the suggestion of Judy, who was a nurse. They were very friendly and wanted to know more about these lectures. I asked Janet, "What is your interest in science?" Janet replied that although she did not enjoy her science classes in school, today she likes learning more about science, especially in the areas of health and the environment. I invited them to come to more of these lectures in the future, which they did! Janet and I took a liking to each other, and in due course we developed a serious relationship, leading to our marriage in September 2015.

21.1 Introduction

Mary, my wife, had died suddenly on December 8, 2011. The next month I was hosting a "Cutting-Edge Lecture in Science" at McGill's Redpath Museum, one of a monthly series of public lectures. Medical professor Alan Evans OC FRS, a brilliant physicist-turned neurophysiologist, was speaking on "Brain imaging: Non-invasive mapping of the brain." Since Mary had died of a hemorrhagic stroke, the lecture certainly held my interest.

After a lively 15-min Q&A period following the lecture, we adjourned to the lobby of the Museum for wine and cheese. Several attendees were keen to have further discussions with Evans, while I circulated in the crowd to welcome newcomers and visit old friends.

With a glass of wine in hand, I introduced myself to Janet and Judy, who were attending this lecture series for the first time. Since Janet posed a question after the talk, I was prompted to ask, "And what is your interest in science?".

> "I hated my science classes in high school," Janet responded. "Now, however, I've become interested in many scientific topics, especially in environment and health."

> "Wonderful," I responded, "I hope this lecture series will nourish your interest in science. I'm glad you came tonight."

L. Mysak, *Adventures in Climate Science, Ocean Waves, and the Flute*, Springer Biographies, https://doi.org/10.1007/978-3-032-19848-8_21

I told Janet that my wife had died very recently, and she in turn shared with me that her husband was in the advanced stages of Alzheimer's disease. But she was occasionally able to go out for an evening and said she really enjoyed the lecture tonight. "Good," I replied, "you and Judy must come to the next lecture in February."

21.2 The Cutting-Edge Lecture Series

The cutting-edge lectures were started in 2003 by McGill professors Graham Bell (an evolutionary biologist and former director of the Museum) and Michael Mackey (a mathematician-turned physiologist). From the get-go, the series was meant to foster communication of recent scientific advances with the public. Each speaker in the series was encouraged to make their presentation jargon-free and accessible to a wide range of people. With typical audiences of 75–100 every month, we felt this outreach program was very successful. After Bell quit as director of the Museum around 2005, I joined Michael as co-director of the lecture series. A few years later, Ingrid Birker joined the Museum staff as Coordinator of Public Education and Outreach. The three of us then worked together to get funding for the series from the McGill administration and to seek out inspiring colleagues to give the lectures. Ingrid also brought on board undergraduate students to help with the series.

Until the economic downturn in 2008–09, we had an annual budget of over $20,000 to support the lecture program. The funds were generously provided by the Principal, Provost, VP Research, and several deans. At this level of funding, we could invite outstanding speakers from across Canada and beyond. After 2008–09, however, our support came mainly from a few deans (especially the Dean of Science), which meant we could invite only local speakers. We also asked the public to make donations to the cutting-edge fund, but these never made up more than 25% of the costs.

21.3 We Meet Again

I was happy to see Janet and Judy back for the February lecture by anthropologist Andre Costopoulos on "A diversity-tolerance model of cultural evolution." (We scientists are often not good at finding audience-friendly titles!) Afterwards we tried to figure out the speaker's take away message, which was not obvious. We think he argued that while humans are smart enough to come up with a range of solutions to the problems we face, we are not very good at determining which solutions are best.

Janet and I also chatted about how we were getting on with our challenging circumstances—her husband's Alzheimer's condition, and my loneliness as a recent widower. Janet was amazed that I remembered many of the details of her personal situation.

Janet and Judy missed the March lecture (on tropical diseases) because of travel away from Montreal but returned for the final lecture of the series in April. John Stix, a vulcanologist, literally exploded with enthusiasm when he spoke about "The nature and origin of large volcanic eruptions." We agreed afterward that this was one of the best public lectures we'd ever heard. Not a surprise. John had won many awards for his outstanding undergraduate teaching at McGill.

In April, with the help of my two children, Paul and Claire, I was also planning a "celebration of life" ceremony for Mary next month. I told them that after this, I would start moving forward to find another partner. I began to invite a few single women out. I also asked Janet and Judy to come for a late spring lunch at my cottage in the Eastern Townships of Quebec (Chap. 29). They were delighted and offered to bring the lunch—which they did! The food was tasty but simple—vichyssoise, chicken salad, and homemade chocolate cookies, which we washed down with a chilled chardonnay.

We had a lovely time that sunny Saturday, along with my friend Hans Zingg, whom I had also invited. After lunch we had a nippy swim in the unheated cottage pool (it was late May), and we strolled in the neighboring forest and farm field. At the end of our walk, I was touched when Janet presented me with a bunch of wildflowers she had picked in the field.

The next day, I had calls of thanks from Janet and Judy. Janet then invited Hans and me to a Rotary Club fund-raising lobster dinner in the Town of Montreal West the following Saturday. Hans and I joined a group of Janet's friends, which included two widowers, Allan Aitken and Bill Ritchie. Ten of us sat around a large table devouring freshly cooked whole lobsters and consuming several bottles of wine. Allan, like me, had lost his wife in the previous year; he looked in a daze. Bill Ritchie was in Montreal for meetings at McGill's Faculty of Agriculture and had a free evening to join us. Janet had asked me to drive him to the Rotary dinner. The conversation at the table was lively and stimulating; I had been missing this type of social activity.

After dinner, I invited Bill, Hans, and Janet to my home for a brandy. I was disappointed when Janet declined. (She had to drive home some of the other party friends.) "Can I take a rain check?" she asked.

21.4 Our Relationship Develops

I didn't know how to respond to this question, but I was certainly attracted to Janet and wanted to see her again. She was lively in conversation and seemed to have unlimited energy. To help me resolve this quandary, I phoned Helen, my sister in Vancouver. She gave what proved to be excellent advice.

The next day I phoned Janet. "I'm not sure if this is appropriate, but if you would like to meet for a coffee or lunch, I'd be happy to do this," I said up front.

"I would love to," she replied. And this is how our relationship started.

We met for lunch at Sakura, my favorite Japanese restaurant near McGill, and afterward we walked on the forested trails of Mount Royal. I then showed her my cluttered office—she was curious to see where I hung out during the day.

Around this time, I told my children that I would like to meet another woman, rather than live the life of an aging widower. They were happy with this, but Claire, then 34, had two concerns. "Daddy, I don't want you to get involved with someone younger than me, or a woman with young children at home," she stated.

"Not a problem," I said. "Not my wish either."

When Janet returned in the fall for the next set of cutting-edge lectures, we continued our friendship. We also made friends with regular attendees of the lectures.

In April 2013, Janet's husband died in a nursing home. After this we spent more time together, and I met her two sons, Ian (in Montreal) and Rob (in Victoria, BC) and their families. Janet and I cycled a lot, enjoyed spending time at the cottage, and I relished her cooking. The latter proved to be key to get the blessing of Aunt Mary, wife my favorite Uncle Orest who lived in Saskatoon. "I'm so happy you've met a special lady," Uncle Orest said to me over the phone one day. "But can she cook?" asked Aunt Mary very bluntly. Mary was renowned for her Ukrainian cuisine, and she was naturally concerned that I should eat well. "Yes, she can, very well," I quickly replied.

21.5 Our Marriage

In the summer of 2015, I sold my Montreal West home and later moved with Janet into an apartment in the Victoria village area of Westmount, QC. In September, we married in the cottage garden (Photo 21.1) in the presence of our immediate families (Photo 21.2). We have had a very active and happy life ever since.

Photo 21.1 Janet and me arriving on bicycles for our wedding in the garden of our country cottage (in Dunham, QC)

21.6 Epilogue

The in-person cutting-edge lectures continued until February 2020 when COVID-19 hit us all. For McGill's bicentennial celebration in 2021, a special cutting-edge set of eight livestreamed lectures was organized by Ingrid Birker and her colleagues for 2021–22. Figure 21.1 lists the speakers for the winter segment of the lectures, which covered both science and social science topics. Ingrid retired in the summer of 2022, and the cutting-edge lecture program ended as there was no one else still working at McGill willing to organize them. I was of course disappointed to see this happen. However, I was very much enriched for 17 years (2003–2020) by meeting many brilliant scientists, keen members of the public, and Janet, who became my wife.

Photo 21.2 The "official" photograph of the blended family after our wedding at the cottage in September 2015. From L to R: Rob Boeckh, Claire Mysak (my daughter), Ian Boeckh (Rob and Ian are the sons of Janet), Jonah Boeckh (son of Rob and Marina), me, Janet, Eric, Sophie and Andrew Boeckh (the three children of Ian and Karen), Karen Boeckh, Marina Steele, Leslie Ingledew (sister of Janet), Greg Osoba (my nephew), and Paul Mysak (my son). In front of Sophie is Calla Boeckh, daughter of Rob and Marina. Missing: Claire's husband Dennis Taylor, and Neil and Austin Talyor, the sons of Claire and Dennis

Cutting Edge Lectures in Science

Les conférences « À la fine pointe »

Bicentennial Winter Program

Programme d'hiver du Bicentenaire

2022

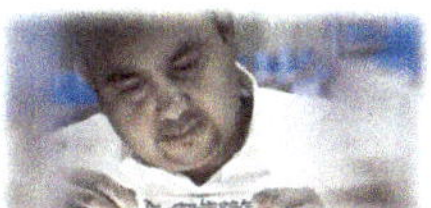

Jan. 13 janvier: Unravelling the secrets of symbiosis - How ants and bacteria became one
By/Par: **Ehab Abouheif** (Dept. Biology / Département de biologie/ l'université McGill University)

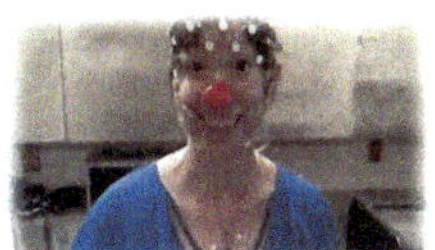

Feb. 10 février: Detecting Consciousness in the Absence of Responsiveness
By/Par: **Stefanie Blain-Moraes** (School of Physical and Occupational Therapy, McGill Faculty of Medicine and Health Sciences / École de physiothérapie et d'ergothérapie, Faculté de médecine et des sciences de la santé de l'Université McGill)

March 10 mars: Combating religious fundamentalism through speculative urbanism in the Gulf
By/Par: **Sarah Moser** (Dept. Geography / Département de géographie, l'université McGill University)

April 14 avril: How brain circuits transduce social information to modulate the learning and control of birdsong.
By/Par **Jon Sakata** (Dept. Biology / Département de biologie/ l'université McGill University)

6 pm / 18 h

Livestreamed with McGill Youtube / Retransmis en direct par McGill Youtube

Fig. 21.1 Cutting-edge poster for the McGill bicentennial winter 2022 program. Figure courtesy of the Faculty of Science, McGill University, Montreal

Acknowledgements I especially thank Janet Boeckh for being my first reader and Josef Schmidt and Mary Feher for their helpful comments. Olivia Marino kindly inserted the photographs. During the last few years of the cutting-edge program, Bruce Lennox (the then Dean of Science) provided a large portion of our funding. Michael, Ingrid, and I are most grateful for his support.

Chapter 22
Going International: My Long Association with IAPSO and IUGG

Abstract In January 1974, I flew to Melbourne, Australia to present, for the first time, a paper at the assembly of the International Association for the Physical Sciences of the Oceans (IAPSO). I was thrilled to meet in Melbourne leading oceanographers from around the world, and after this meeting I made a vow to attend the biannual IAPSO assemblies in the future. In 1975 I presented another paper at the general assembly of the International Union of Geodesy and Geophysics (IUGG) in Grenoble. Every four years, IUGG brings together the eight sister associations of earth and space science that comprise IUGG. After attending these assemblies for three decades, I was invited to join the IAPSO executive as a vice president in 2003. It was time to "give back" and take my turn to help organize these international conferences in the future. I served as president of IAPSO and a vice president of IUGG during 2007–2011, which was both a rewarding and challenging experience. I enjoyed developing friendships and collaborations with scientists from many countries, and I worked hard to establish programs and projects which brought together oceanographers and climate scientists from all continents.

22.1 Introduction

Paola Rizzoli, a distinguished MIT professor of oceanography originally from Italy, is a very persuasive woman. As president of IAPSO (International Association for the Physical Sciences of the Oceans), Paola phoned me in the spring of 2003. "Lawrence, I want you to join the IAPSO executive as a vice president. If you agree, this will be made official this summer at our meeting in Sapporo, Japan, where I know you'll be giving a couple of papers."

"Why me?" I asked. I had been participating in international meetings of IAPSO since 1974, but I had no aspirations of taking on any leadership roles in this organization. In fact, at the ripe age of 63, I was already planning for my retirement, although I must admit that I wanted to continue presenting papers of my latest research at future IAPSO meetings and meet colleagues and friends from around the world. This has always been an important part of my scientific career.

L. Mysak, *Adventures in Climate Science, Ocean Waves, and the Flute*, Springer Biographies, https://doi.org/10.1007/978-3-032-19848-8_22

"You did an excellent job of organizing a two-day symposium on oceanography and climate of the Arctic at our meeting in Mar del Plata (Argentina) in October 2001. We need to bring in new blood into the leadership of our oceanography association," she replied. "You can't say no!"

In July 2003, I was duly elected vice president of IAPSO in Sapporo, where the oceanography association was taking part in the 23rd General Assembly of IUGG (International Union of Geodesy and Geophysics). This was the first time IUGG had met in Asia, and I soon realized that I had much to learn about the workings of IAPSO and IUGG.

IUGG was founded in Brussels as a nongovernmental organization (NGO) in 1919 by nine countries (including Canada) under the auspices of the International Research Council (now the International Science Council, ISC). At the time of writing this chapter in 2024, there were 74 regular member countries of IUGG and a small number of affiliates. As stated on its website, "IUGG is dedicated to the international promotion and coordination of scientific studies of the Earth and its environment in space." In addition to IAPSO, IUGG has seven other semi-autonomous associations representing (1) atmospheric sciences, (2) hydrology, (3) the cryosphere (science of ice), (4) seismology, (5) vulcanology, (6) geodesy (the study of gravity and the shape of the earth), and (7) geomagnetism and aeronomy. The first meeting of IUGG took place in Rome in 1922, and thereafter IUGG met every three years in different European and North American cities until 1963.

Starting in 1967, the general assemblies of IUGG were held every four years, and in intervening years the associations started meeting on their own, the first one being the IAPSO assembly in Tokyo in 1970. In January 1974, I attended my first meeting of IAPSO in Melbourne, which was a joint assembly with International Association for Meteorology and Atmospheric Physics (IAMAP, now the International Association for Meteorology and Atmospheric Sciences, IAMAS). The trip to Melbourne had the bonus that it was my first return to Australia since 1962 when I was a graduate student in Adelaide (Chap. 5). Also, in Melbourne my UBC colleague Paul LeBlond (Photo 22.1) and I signed a contract with Elsevier Scientific Publishers to write a graduate level textbook on ocean waves (Chap. 13).

In September 1975, I attended my first IUGG general assembly in Grenoble, which proved to be most helpful when writing *Waves in the Ocean* [1]. At this conference I learned, for example, that my draft chapter on equatorially trapped waves was badly out of date. This took me back to the drawing board when I returned to Vancouver in the fall. In the revised chapter, I now showed, for example, that equatorial waves play an important role in the Southwest Monsoon of the Indian Ocean and the generation of the Somali Current [2].

Photo 22.1 Paul LeBlond (on the right) with other oceanographers in Melbourne, during the 1974 IAPSO assembly. From left to right: Ted Buchwald (Australia), George Needler (Canada), Claude Frankignoule (France), Anatoli Volochkov (Russia), Boris Gavrilin (Russia), and LeBlond (Canada)

22.2 My Term as Vice President of IAPSO, 2003–07

My first executive meeting of IAPSO in Sapporo was a bizarre experience. Japanese oceanographer Shiro Imawaki was the incoming president of IAPSO, and his command of English was limited. When chairing his first executive meeting, he was continually being interrupted by other executive members, including a couple of former IAPSO presidents who should not have been there. I felt sorry for Shiro. As I did not know many people at the meeting and was new to the executive, I couldn't find an easy way to help him out. Fortunately, the situation improved when we started talking about specific symposia topics being proposed for the next IAPSO meeting, which was going to take place in Cairns, Australia, in August 2005. Now, Shiro could focus on science and not have to deal with vague mission statements which other executive members kept bringing up.

During my term as vice president, I learned that SCOR, the Scientific Committee on Oceanic Research (another NGO under the umbrella of ISC), had an important liaison with IAPSO. SCOR was founded in 1957 with the aim of promoting international cooperation in planning and conducting research in all areas of oceanography. Since IAPSO was concerned with the mathematical, physical, and chemical aspects of oceanography, this liaison with SCOR permitted us to develop new working groups and research programs on ocean topics which brought in the biological and geological aspects of oceanography.

In Sapporo, Paola proposed, "Why don't you organize an IAPSO meeting in Montreal?" Since there was already one planned for Cairns in 2005, she was thinking of 2009! Long range planning, indeed.

"Good idea," I replied, "and we should also include the atmospheric people". Since I was now doing research in atmosphere–ocean interactions and their role in climate variability and change [3–5], I thought it would be nice to have a joint assembly of the ocean and atmosphere associations of IUGG. Paola subsequently introduced me to Mike MacCracken, the then president of IAMAS. When I proposed this to him, he replied, "Yes, we should think about this, and I'd be in favor if you took the lead on organizing such a meeting." Ouch, I thought, this sounds like a lot of work.

At the end of the IUGG meeting in Sapporo, we learned that Perugia, Italy, was selected as the site of the 2007 General Assembly of IUGG. Rizzoli was very happy with this choice and started describing the beauty of Perugia and its location on top of an extinct volcano. Imawaki would end his term there as president of IAPSO, and we would need to elect a new president. At this time, we also knew that Tom Beer, a friend and atmospheric scientist from Melbourne, Australia, would become the next president of IUGG for the period 2007–11. I thought that he would make an excellent president.

In August 2005, I did attend the next meeting of IAPSO in Cairns and learned more about the activities of our association, especially the working groups we organized with SCOR. A SCOR Working Group (WG) brings together about a dozen leading oceanographers from around the world to address scientific issues (both conceptual and methodological) that are impeding the advancement of ocean science. At the time, IAPSO and SCOR jointly sponsored the following working groups: Ocean Mixing (WG 121), Mechanisms of Sediment Retention in Estuaries (WG122), Thermodynamics and Equation of State of Seawater (WG127), Deep Ocean Exchange with the Shelf (WG129), and OceanScope (WG 133). I was very impressed with this wide range of activities that IAPSO was involved in.

The trip to Cairns was also special for me personally. Enroute to this city, my then wife Mary and I spent a wonderful week in New Zealand thanks to my Kiwi friend and oceanography colleague Malcolm Bowman (Photo 22.2). Later, while in Cairns, we took a spectacular tour of the Great Barrier Reef (Photo 22.3). At the IAPSO meeting itself, Fritz Scott from Germany was awarded the 2005 Prince Albert I Medal, an IAPSO biannual award set up by Rizzoli in 2001 to recognize a prominent physical or chemical oceanographer. The selection committee for the medal is chaired by a vice president of IAPSO, and nominations are sought from all member countries of IUGG. Prince Albert I (1848–1922), who devoted much of his life to oceanography, exploration, and science, organized the physical oceanography section of IUGG at its founding in 1919. This section evolved into IAPSO in 1967.

Fritz Schott was a notable sea-going oceanographer who reached out to theoreticians like me who used mathematical models to explain oceanic current observations. During a summer visit to UBC in 1976, Fritz showed me some puzzling sub-inertial current fluctuations that he observed in the Norwegian Current. He and

Photo 22.2 Left: The author sharing a view of Auckland harbour with Malcolm Bowman (on my left), now emeritus professor of oceanography at State University of New York, Stony Brook. Right: With my then wife Mary Mysak on Kane Beach (near Auckland), the site for the movie *Piano*

Photo 22.3 Fish and coral seen through a glass-bottomed boat in the Great Barrier Reef of Australia. (Photograph taken by Mary Mysak in 2005.)

Photo 22.4 On the left, IAPSO past president Paola Rizzoli and president Shiro Imawaki at a dinner in Cairns, August 2005. I'm in a dark suit on the right with other delegates at the conference

I were delighted when a simple baroclinic instability model of a two-layer flow in a channel could be used to explain these observations [6]. Sadly, Fritz died in 2008, three years after receiving the prestigious Prince Albert I Medal.

While in Cairns, Rizzoli was also pressuring me to be a candidate for the next president of IAPSO (Photo 22.4). "You have many contacts in the oceanographic and climate science community, and you're widely recognized for your research," she said. "You would make an excellent president." Somewhat reluctantly, I agreed to be a candidate. As I said at the beginning of this chapter, Paula is a very persuasive woman. In Perugia I was indeed elected president and had the honor of being the second Canadian to hold this position. Robert W. Stewart OC FRS, an ocean turbulence expert from British Columbia, was IAPSO president during 1979–83. Professor Johan Rodhe from Sweden was elected Secretary General of IAPSO. I soon learned that Rodhe would play a crucial and helpful role in the operations and activities of our association.

22.3 My Term as President of IAPSO, 2007–11

One of the challenges of IAPSO is to make the organization relevant to physical and chemical oceanographers from around the world. Each member country of IUGG has a national committee which includes representatives from each association. These representatives are invited to the associations' biannual general business meetings to learn about the activities of the association, such as the SCOR-IAPSO working groups mentioned above. However, I felt that more frequent communication with the IAPSO representatives was needed. This prompted me to write a semi-annual IAPSO letter to these representatives and to prepare a brochure which described the objectives and activities of IAPSO (Fig. 22.1). To carry out these projects, I had the able assistance of IAPSO secretary Johan Rodhe (Photo 22.5).

To make the IAPSO general business meetings more engaging, I asked the chairs of SCOR-IAPSO WGs to give progress reports on their work, which happily led to active Q&A sessions afterward. We also had informative reports from the heads of IAPSO Commissions and Services, which included Mean Sea Level and Tides, Tsunamis, and Seawater. We also served tea and cookies before and after these meetings, which helped produce a collegial atmosphere among the national delegates. This prompted Robin Muench, a former president of IAPSO (1991–95), to say at the 2011 IAPSO meeting in Melbourne, "Lawrence, this is one of the best run business meetings of IAPSO that I have ever attended."

IAPSO

The International Association for the Physical Sciences of the Oceans

http://iapso.iugg.org

For more information contact the Secretary General:

Professor Johan Rodhe
Department of Earth Sciences
University of Gothenburg
PO Box 460
SE-405 30 Göteborg
Sweden
Tel: +46 31-786 28 76
Fax: +46 31-786 49 03
E-mail: johan.rodhe@gu.se

President:

Dr. Lawrence A. Mysak
Canada Steamship Lines Professor
Dept. of Atmospheric and Oceanic Sciences
McGill University
805 Sherbrooke St. W.
Montréal, QC H3A 2K6
Canada
Tel: 514-398-3768
Fax: 514-398-6115
E-mail: lawrence.mysak@mcgill.ca
Website: www.esmg.mcgill.ca

The International Association for the Physical Sciences of the Oceans

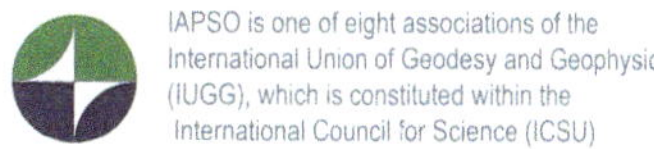

Fig. 22.1 The IAPSO brochure prepared during my time as president, which was produced by the IAPSO Secretary General Johan Rodhe. Permission granted by the current Secretary General of IAPSO, Sylvia Blanc

Photo 22.5 On the far right, Swedish oceanographer Johan Rodhe, Secretary General of IAPSO (2007–2015) in Perugia, Italy, July 2007. Others in the photo (left to right): Me (as new IAPSO president), Shiro Imawaki (outgoing IAPSO president), and his wife Junko, next to Rodhe

As IAPSO president, I was invited to the annual SCOR planning and information meetings. This took me to Bergen, Norway (2007), Woods Hole, Massachusetts (2008), Beijing (2009), and Toulouse, France (2010), where I heard up-to-date reports on international oceanographic programs and new scientific discoveries. In Toulouse, I was especially impressed by the recent work on marine biodiversity [7].

In 2003, IAPSO created the Eugene Lafond Medal for emerging oceanographic researchers from developing countries. Lafond was a long-time Secretary General of IAPSO (1970–87) who made great efforts to promote ocean studies and research in Africa, South America, and Asia. During my term as president, we chose winners of this medal by the quality and clarity of their presentations at an IAPSO assembly. The medal was awarded as a highlight of the final business meeting of the IAPSO assembly.

During the first two years of my presidency, I spent a lot of time helping to organize the 2009 joint assembly of IAMAS, IAPSO, and the newly created IACS in Montreal. At the 2007 IUGG General Assembly in Perugia, the International Association for Cryospheric Sciences (IACS) was spun off from IUGG's IAHS (International Association of Hydrological Sciences). Since the cryosphere plays an important role in climate change, both past and future, it was natural that we invite this new association to join the oceanographers and meteorologists in this assembly. The meeting soon had the nickname MOCA, for "Meteorology, Oceanography and

Cryosphere Assembly". With over 1300 delegates, including IUGG president Tom Beer, the assembly was a great success. Harry Bryden FRS, a colleague now based in England, was awarded the 2009 Prince Albert I Medal in Montreal. In his medal acceptance speech, Harry mentioned that he had applied to UBC to be my graduate student in theoretical oceanography in the early 1970s. However, he decided to stay in the USA to do a PhD in oceanography at MIT. He worked at the Woods Hole Oceanographic Institute (WHOI) for over a decade before moving to Southampton in the UK in the 1990s. Nevertheless, we stayed in touch, and over the years I heard many of his excellent talks at various IUGG general assemblies. I am delighted that I ended up writing my last scientific publication with Harry as a coauthor [8].

A major event during my presidency was the release of TEOS-10 (Thermodynamic Equation of Seawater), a manual that is the new international standard for the calculation and modelling of the thermodynamic properties of sea water, humid air, and ice [9]. This manual is the final product of the above mentioned SCOR-IAPSO WG 127, which was chaired by the renowned Australian oceanographer Trevor McDougall FRS. TEOS-10 replaces the less rigorous (i.e., empirical) EOS-80, the former official description of sea water properties. TEOS-10 uses Absolute Salinity S_A (mass fraction of salt in sea water) as opposed to Practical Salinity S_P (a measure of conductivity of sea water) to describe the salt content of sea water. For his leadership and outstanding work in this field, Trevor was awarded the 2011 Prince Albert I Medal at the 25th IUGG General Assembly in Melbourne.

22.4 My Term as Past President of IAPSO, 2011–15

My main task as past president was to mentor the new president, Eugene Morozov from Russia, and to work with the nominating committee to bring new members into the executive of IAPSO in Prague, the location of the 26th IUGG General Assembly in 2015. At the time, the IAPSO executive consisted of 12 members: president, past president, two vice presidents, secretary general, treasurer, and six members-at-large. Denise Smyth-Wright, a chemical oceanographer from the UK, became the next president in 2015. (Denise, in a white sleeveless dress, is sitting to my left in Photo 22.4) Denise had served on the IAPSO executive as member-at-large for many years and would wisely use this experience to lead our association. The new Secretary General and Treasurer were respectively, Stefania Sparnocchia from Italy and Ken Ridgway from Australia. The incoming vice presidents were Trevor McDougall (Australia) and Isabelle Ansorge, from South Africa. The six executive members-at-large spanned the world and came from Sweden, the Netherlands, Japan, Puerto Rico, the USA, and India. Five of the twelve were women oceanographers.

I fondly remember the IUGG General Assembly in Prague. We had sunny weather, excellent food, and generous servings of wine at the evening poster sessions, held right after the day-time oral presentations. It was also a happy time for me personally as I was accompanied by my fiancée, Janet Boeckh (Photo 22.6), who had relatives (through her previous marriage) living in Prague. We cycled with them in the lush

Bohemian forests surrounding Prague. At the closing ceremony, I was recognized for my contributions to IAPSO and IUGG with a certificate of appreciation and Fellowship in IUGG (Fig. 22.2). We also learned in Prague that Montreal would host the 27th IUGG General Assembly in 2019.

Photo 22.6 Cruising on the Vlatava River in Prague, July 2015 with Janet Boeckh. We got married in September

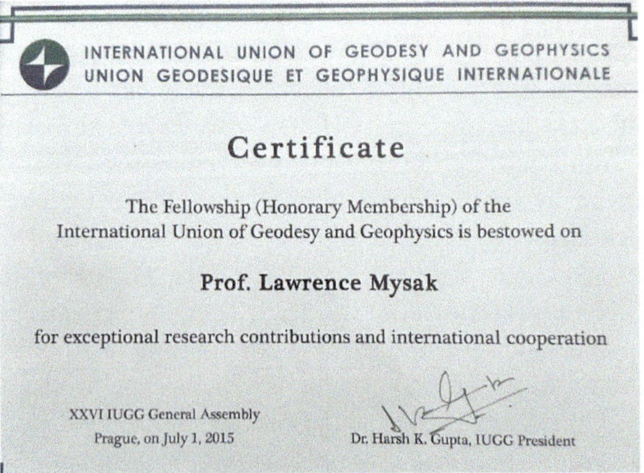

INTERNATIONAL UNION OF GEODESY AND GEOPHYSICS
UNION GEODESIQUE ET GEOPHYSIQUE INTERNATIONALE

Certificate

The Fellowship (Honorary Membership) of the International Union of Geodesy and Geophysics is bestowed on

Prof. Lawrence Mysak

for exceptional research contributions and international cooperation

XXVI IUGG General Assembly
Prague, on July 1, 2015

Dr. Harsh K. Gupta, IUGG President

Fig. 22.2 Certificate of Fellowship in IUGG

22.5 Epilogue

During the 45 years that I have participated in IAPSO and IUGG assemblies, I have presented over 30 papers and have developed friendships and research collaborations with many colleagues, as have my graduate and postdoctoral students. Some of my most interesting papers have come about because of chance conversations during coffee breaks.

At the 1979 IUGG General Assembly in Canberra, Australia, Robin Muench from Seattle suggested that a model I used to explain sea surface temperature patterns in the California Current off Vancouver Island might also be used to explain current fluctuations he observed in Shelikof Strait, Alaska [10]. In Vancouver, at the 1987 IUGG meeting, I had exciting discussions with European oceanographers about the "Great Salinity Anomaly" (GSA) which was observed in the Greenland and Labrador seas in the late 1960s and 1970s [11]. I had just began studying sea ice cover anomalies in the Arctic after my move to McGill in 1986 and wondered whether there was any connection between these and the GSA. This resulted in one of my most cited papers which invokes the concept of a feedback loop to explain natural climate variability on the decadal timescale in the Arctic [12]. This feedback loop was later simplified by Mysak and Venegas [13]. Shortly after this work, I developed another feedback loop to explain the interdecadal variability of northeast Pacific coastal freshwater discharge [4]. Finally, at the 1999 Birmingham IUGG meeting, my friend and colleague André Berger MAE told me about a new class of climate models that was under development in Europe and elsewhere. My research group at McGill was able to participate in subsequent European workshops on this topic organized by Martin Claussen MAE, and several of us became authors of a paper [14]on Earth system Models of Intermediate Complexity (EMICs). This paper has been widely cited in the recent scientific publications of IPCC (Intergovernmental Panel on Climate Change).

My last duty with IUGG involved my serving as deputy chair (2017–19) of the organizing committee for the 2019 IUGG General Assembly in Montreal. The committee met monthly at UQAM since the chair, Fiona Darbyshire, was a professor of geophysics there. We were happy that nearly 4000 delegates attended the meeting, and my wife Janet and I were able to greet many old friends and colleagues at the opening reception. Unfortunately, toward the end of the conference I came down with a bad cough which developed into pneumonia. However, after an emergency visit to the country hospital where we were holidaying right after the conference, I was able to get the necessary antibiotics that enabled me to recover from this nasty infection.

Before I became ill, I was able to attend the business meetings of IAPSO in Montreal, now being ably chaired by President Denise Smyth-Wright. I was delighted to learn about some of the new directions that IAPSO was taking. These included best practice study groups and early career networks, which has attracted many new young oceanographers to partake in the activities of IAPSO. This was healthy sign for the future of IAPSO which had been dear to my scientific life for nearly five decades.

Acknowledgements I thank Janet Boeckh for her early review of this memoir and Josef Schmidt for his incisive comments. Agatha de Boer, Jacques Derome, Shiro Imawaki, Paola Rizzoli, and Hans van Haren have provided valuable input. I appreciate the help of Olivia Marino for inserting the photographs.

References

1. LeBlond PH, Mysak LA (1978) Waves in the Ocean. Elsevier Scientific Publishing Co, Amsterdam
2. Cox MD (1976) Equatorially trapped waves and the generation of the Somali Current. Deep-Sea Res 23:1139–1152
3. Peng S, Mysak LA, Ritchie H, Derome J, Dugas B (1995) On the differences between early and midwinter atmospheric responses to sea surface temperature anomalies in the northwest Atlantic. J Climate 8:137–157
4. Royer TC, Grosch CE, Mysak LA (2001) Interdecadal variability of Northeast Pacific coastal freshwater and its implications for biological productivity. Progr Oceanogr 49:95–111
5. Venegas SA, Mysak LA, Straub DN (1997) Atmosphere-ocean coupled variability in the South Atlantic. J Climate 10:2904–2920
6. Mysak LA, Schott F (1977) Evidence for baroclinic instability of the Norwegian Current. J Geophys Res 82:2087–2095
7. Costello MJ, Coll M, Danovaro R et al (2010) A census of marine biodiversity knowledge, resources, and future challenges. PLoS ONE 5(8):e12110
8. Bryden HL, Mysak LA (2018) Ocean circulation: knowns and unknowns. In: Beer T, Li J, Alverson K (eds) Global change and future earth: the geoscience perspective, chap 12. Cambridge University Press, pp 159–175
9. IOC, SCOR, IAPSO (2010) The international thermodynamic equation of seawater—2010: Calculation and use of thermodynamic properties. Intergovernmental oceanographic commission, manuals and guides No. 56, UNESCO (English), pp 196
10. Mysak LA, Muench RD, Schumacher JD (1981) Baroclinic instability in a downstream varying channel: Shelikof Strait, Alaska. J Phys Oceanogr 11:950–969
11. Dickson RR, Meincke J, Malmberg SA, Lee JA (1988) The "Great Salinity Anomaly" in the northern North Atlantic 1968–1982. Progr Oceanogr 20:103–151
12. Mysak LA, Manak DK, Marsden RF (1990) Sea-ice anomalies observed in the Greenland and Labrador seas during 1901–1984 and their relation to an interdecadal Arctic climate cycle. Clim Dyn 5:111–133
13. Mysak LA, Venegas SA (1998) Decadal climate oscillations in the Arctic: A new feedback loop for atmosphere-ice-ocean interactions. Geophys Res Lett 25:3607–3610
14. Claussen M, Mysak LA, Weaver AJ, Crucifix M, Fichefet T, Loutre M-F, Weber SL, Alcamo J, Alexeev VA, Berger A, Calov R, Ganopolski A, Goose H, Lohmann G, Lunkeit F, Mokhov II, Petoukhov V, Stone P, Wang Z (2002) Earth system models of intermediate complexity: closing the gap in the spectrum of climate system models. Clim Dyn 18:579–586

Chapter 23
Stained Glass, Outdoor Hockey Rinks, and Climate Change

Abstract This chapter gives an overview of two projects by McGill MSc students dealing with the impacts of climate change. In the first project, the research shows that as the climate in Europe shifted from the medieval warm and sunny period to the cool and cloudy Little Ice Age period, the stained glass in northwestern European churches changed from dark colors to paler or grisaille (clear) colors. In the second project, reflecting a Canada-wide passion for ice hockey, the length of the outdoor skating season in large parts of Canada is shown to have shortened by up to two weeks over the past half century in response to global warming. Assuming this warming trend continues, our cherished backyard hockey rinks will disappear in a few decades, and future Canadian hockey players will have to get their start on artificial ice rinks.

23.1 Stained Glass Church Windows in Western Europe

Since the 1980s, my research group at McGill has focused on trying to understand natural (internal) climate variability and human-induced climate change using Earth system models and analyzing large atmospheric, oceanic, and sea ice data sets. I never expected that I would supervise research on the transmissive properties of stained glass church windows and the natural interior lighting in Gothic and Renaissance churches in western Europe [1, 2]. How did this come about?

When I was Graduate Program Director in McGill's Department of Atmospheric and Oceanic Sciences (AOS) in the 2000s, I had the responsibility of ensuring that every graduate student had an appropriate supervisor for his or her MSc or PhD thesis research. In spring 2007, MSc student Chris Simmons had just finished his graduate courses with a straight-A record and was searching for a thesis supervisor. He asked, "Can you please suggest who in the department might be interested in helping me establish a connection between the properties of stained glass windows in European churches and climate change?" I didn't have a clue what he was talking about. "Tell me more, please," I replied.

He then went on to argue that during the Medieval Warm Period (eleventh to the end of the thirteenth century), when the atmospheric conditions were warm and

L. Mysak, *Adventures in Climate Science, Ocean Waves, and the Flute*, Springer Biographies, https://doi.org/10.1007/978-3-032-19848-8_23

sunny in western Europe (western Germany and northern France in particular), the Gothic churches in this region tended to have a lot of dark stained-glass windows, especially ones with a deep red color. However, during the Little Ice Age (fourteenth and fifteenth centuries and beyond), when the weather conditions were cooler and often cloudy, the church windows became larger and paler in color (e.g., grayish) or clear (grisailled), which allowed larger amounts of light into the church interior. He wondered if this was just an accident or whether the church masons were deliberately responding to the slow change in climate, from the warm and sunny Medieval Warm Period to the cooler and cloudy Little Ice Age.

"But what research will you do?" I asked. "I will first make light transmissivity measurements of the stained glass windows in these churches using a digital camera," he replied. "Then I will collect luminance data from the interior of these churches, to see how the interior lighting would have changed over the centuries."

As I did not know of anyone in the department who could be interested in this sort of research, I hesitantly agreed to be his MSc thesis supervisor. I could see that Chris was keen to work on this project, and that he could do it quite independently. Fortunately, I knew McGill architecture professor David Covo who in turn introduced us to Dr. Jennifer Veitch from Ottawa's Institute for Research in Construction, a part of the National Research Council of Canada. She and her colleagues at NRC provided us with valuable advice on the technical execution of the research (e.g., she advised us on the types of light measurements to make and with what equipment). With this help, Chris, his mother Teresa, grandfather Robert and Chris's friend Cathérine Thauer set off to western Europe in summer 2007 and winter 2007–8 to collect data on the transmission properties of the windows and the amount of natural daylighting in a dozen cathedrals and churches in France, Germany and Spain (Table 3 in [1]). Included in this table, for example, are Chartres Cathedral (France), Cologne Cathedral (Germany), and a church in Toledo (Spain). The overarching rationale for this study is that lighting, architecture, and climate have always been linked together in the design of churches during the Gothic and Renaissance periods [3].

Chris and his team used high dynamic range (HDR) imagery (a low-budget, time-efficient means of photographic data collection) to estimate the relative transmissivities of adjacent panels in stained glass windows in the churches listed in [1]. They found that white and yellow-dominated stained glass from the fourteenth and fifteenth century churches admits 5 to 10 times more light into the church interior than the full-color (dark red, green, and blue) windows of the thirteenth century churches. This result is consistent with the hypothesis that during the Little Ice Age, when the conditions were cloudier, the church windows needed to have whitish windows to let sufficient light in.

In a follow-up paper [2], Chris and his team used four illuminance meters (see [2, 4] for details) to measure the natural daylighting in the churches during sunny and cloudy conditions. Remarkably, they found that during sunny conditions, the difference in interior lighting between dark-color windows and grisailles (whitish or clear windows) is relatively small. However, during cloudy conditions, there is a broad fivefold increase in lighting for the grisailled windows used in the fourteenth and fifteenth centuries. This difference is especially notable in winter when there is generally increased cloudiness.

The long-term change in cloudiness most likely occurred because of the shift in the North Atlantic Oscillation index from positive (deep Icelandic Low) to negative (weak Icelandic Low) values around the beginning of the 1400s (see Fig. 6 in [2]). When this happens the weather conditions over western Europe tend to be unstable and cloudier.

23.2 Outdoor Hockey Rinks in Canada

In the spring of 2010, AOS MSc student Nikolay Damyanov met Chris Simmons who was now doing a PhD at McGill on modeling the natural carbon cycle during the Holocene [5] under the joint supervision of myself and Damon Matthews from Concordia University. Nikolay was intrigued by the stained glass project of Chris. "Is there another 'real world' climate-impact study that I could do?" he asked. "I'm not interested in doing any modelling work." Since I was about to retire formally on 1 June 2010, I told Nikolay that after this date I would not be allowed to supervise alone any new students, but I said that I would talk to Damon about the possibility of co-supervising a climate-impact project. Damon was an adjunct professor at McGill who could serve as an official co-supervisor, and I knew he had a student investigating winter temperature change in Montreal.

Between 1950 and 2005, winter temperatures across Canada had increased over 2.5 °C, far above the globally averaged increase that is attributed to global warming [6]. Damon wondered whether these warmer temperatures had any noticeable impact on the quality of outdoor skating rinks in Canada over this period. After conversations with Mike Hudema, Shawn Marshal and Sarah Turner, Damon decided to focus on determining for the above period the annual start of the outdoor skating season (OSS), and how long the OSS lasted for different cities and regions of Canada (as defined in Fig. 23.1). Nikolay was interested in working on this project.

Nikolay's first task was to consult officials responsible for maintaining public outdoor skating rinks in Toronto and various cities on Montreal Island about starting and end dates for the OSS. From discussions with these officials, he developed a meteorological criterion to mark the beginning of the OSS at a given station, namely, the last day in a series of the first three consecutive fall/winter days with a maximum air temperature below – 5 °C. This criterion is based on the requirement of several consecutive cold days to lay the initial ice foundation of a rink on a frozen ground.

Nikolay found it more of a challenge to find the end date for the OSS at a given location because the ice-breakup is not easily related to the daily maximum temperature. Furthermore, many cities stop maintaining rinks after the end of February. After much consideration, he defined a proxy for OSS length as the total number of days with a maximum temperature below – 5 °C after the OSS start date, and before the beginning of March (see [7] for further information).

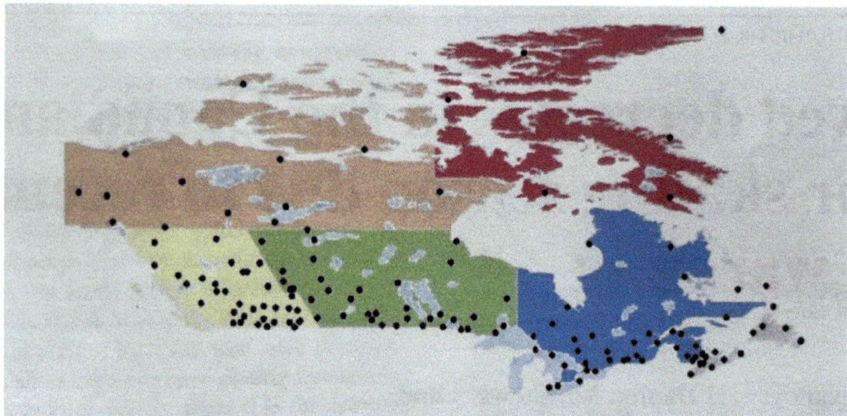

Fig. 23.1 Locations of the 142 meteorological stations and six climatic regions used in [7]. Western regions include the Southwest (yellow), Prairies (green), and Northwest (brown). Eastern regions are Central Canada—Ontario and Quebec (blue), Atlantic Canada (grey) and the Northeast (red). Figure from Damyanov et al. [7]

Nikolay then applied the above criteria to a 55-year surface air temperature data set (1951–2005) developed by the Meteorological Service of Canada (see [7] for details) to determine the annual OSS start date and length for each station shown in Fig. 23.1. The trends for these quantities at each station were also computed (Fig. 2 in [7]). Nikolay then computed the regional trends by spatially averaging the annual OSS start dates and lengths for each station within each region shown in Fig. 23.1. Superimposed on these regional trends is considerable interannual variability (Fig. 3 in [7]). For the three western regions, much of this variability can be attributed to ENSO events, whereas for the three eastern regions, the variability is generally associated with the NAO. However, an inspection of the regional trends reveals that most regions across Canada had seen a significant decrease in the length of the OSS during 1951–2005. The largest decreases (about two weeks) occurred in the Prairies and Southwest Canada.

Damyanov et al. [7] also made some projections for OSS lengths under continued winter climate warming. In the most extreme case of Southwest Canada, a simple linear extrapolation of the OSS length trend from the last 30 years of our record into the future shows that the number of viable rink-flooding days could shrink to zero by 2050. Not good news for future generations of young children who want to emulate former Edmonton Oiler hockey star Wayne Gretzky who learned to skate and play hockey in a backyard rink maintained by his father.

23.3 Epilogue

Since the publication of our climate-impact papers [1, 2, 7], Google Scholar reports (April 2026) a total of 54 citations for them, with 64% being for the outdoor skating rink study of Damyanov et al. Although the Atmosphere–Ocean paper [2] has had the least number of citations, I find this the most intriguing one of the three listed above. However, since the publication of the outdoor skating paper and the story of this study in the McGill Reporter in 2012 (Photo 23.1), I have had many calls, even in 2025, from newspapers and TV stations about the future of outdoor skating rinks (and cross-country skiing) in Quebec. And the same is true for Damon Matthews. In my own case, I remember that in the 1980s and 1990s I used to start cross-country skiing in Montreal and the Eastern Townships of Quebec in late November or early December. Today, in the mid-2020s, I rarely start skiing before Christmas, and during January and February, I'm lucky to have half a dozen days of good skiing. In a few years, I may be hanging up my ski poles for good!

Since the OSS publication [7], other universities have been recording the number of outdoor skating days on natural rinks. For example, the science-citizen program Rink Watch at Wilfred Laurier University (Waterloo, ON) tracks data from over 1400

Photo 23.1 The author testing the ice conditions (ha ha) on the walkway near Burnside Hall of McGill University during a thaw in March 2012 (Photo courtesy of Owen Egan; this photo appeared in the 15 March 2012 issue of the McGill Reporter)

outdoor rinks in North America [8]. Thus, the public interest in the OSS in Canada and the USA continues to be strong.

Acknowledgements I thank Janet Boeckh, Josef Schmidt, Nikolay Damyanov, and Damon Mathews for their helpful comments. Claire Mysak's assistance with the figures is gratefully acknowledged.

References

1. Simmons CT, Mysak LA (2010) The transmissive properties of medieval and renaissance stained glass in European churches. Arch Sci Rev 53:251–274
2. Simmons CT, Mysak LA (2012) Stained glass and climate change: how are they connected? Atmos-Ocean 50:219–240
3. Jason A (2001) History of Art, 5th edn. Harry N. Abrams, New York
4. Simmons CT (2008) Fiat lux: climatic considerations in medieval stained glass aesthetics. MSc thesis, McGill University, Montreal
5. Simmons CT, Mysak LA, Matthews HD (2013) Investigation of the natural carbon cycle since 6000 BC using an intermediate complexity model: the role of southern ocean ventilation and marine ice shelves. Atmos-Ocean 51(2):187–212. https://doi.org/10.1080/07055900.2013.773880
6. Hansen J, Sato M, Ruedy R, Lo K, Lea DW, Medina-Elizade M (2006) Global temperature change. Proc Natl Acad Sci USA 103:14288–14293
7. Damyanov NN, Mathews HD, Mysak LA (2012) Observed decreases in the Canadian outdoor skating season due to recent winter warming. Environ Res Lett 7:8. https://doi.org/10.1088/1748-9326/7/1/014028
8. Robertson C, McLeman R, Lawrence H (2015) Winters too warm to skate? Citizen-science reported variability in availability of outdoor skating in Canada. Can Geographies 49:383–390

Chapter 24
Some Graduate Students and Postdocs I've Known

Abstract In this chapter, I profile ten graduate students and five postdocs whom I have supervised or co-supervised during my five decades as a professor, first at UBC and then at McGill. Each student has come under my supervision in a different way, and after leaving UBC or McGill each has generally kept in close contact with me. It has given me enormous pleasure to see how these students have developed in their careers and become leaders in their fields.

24.1 Introduction

An important and rewarding part of the work of most professors at research-intensive universities is the supervision of graduate and postdoctoral students. For 50 years, from my start at UBC in 1967 to the graduation of my last PhD student from McGill in 2017, I have had the privilege of mentoring 80 of these "academic children." The relationship between professor and student is deeply symbiotic. Students need guidance to complete their theses and launch their careers, and professors rely on graduate students and postdocs to help advance their research. In my case students have also helped me with programming, data analysis, and the preparation of conference presentations. What are the factors that contribute to a successful relationship? How are research topics chosen? And which ones lead to breakthrough results? In this chapter, I explore these questions by profiling 15 of my grad students and postdocs, a few of whom can be seen in Photo 29.7. Each came to me in a unique way, and their stories—as theorists, modelers, policymakers, data analysists, and artists—highlight the surprising and fulfilling paths a life in science can take. It has been one of my greatest pleasures to see them develop into leaders in their fields in several countries and to maintain lasting friendships with so many of them.

While some of my students arrived with meritorious graduate or postdoctoral fellowships which covered their tuition and living costs, many of them needed financial support in the way of research assistantships which I provided. These were paid by the peer-reviewed research grants and contracts that I was awarded by various agencies. In addition, these grants and contracts covered the costs of the

L. Mysak, *Adventures in Climate Science, Ocean Waves, and the Flute*, Springer Biographies, https://doi.org/10.1007/978-3-032-19848-8_24

students' and my own computers and other equipment, publications, and conference trips. I was fortunate to have had an uninterrupted sequence of operating grants awarded from NRC (1967–78) and then NSERC (1978–2014; in the early 1980s, the operating grants were renamed "Discovery Grants"). Further federal support came from NSERC Strategic, Equipment, and Collaborative grants, the NSERC Industrial Research Chair program, the Canadian Atmospheric Environment Service, Fisheries and Oceans Canada, the Canadian Institute for Climate Studies, and the Canadian Foundation for Climate and Atmospheric Sciences. I also had a long-standing contract (1979–1992) with the US Office of Naval Research (working on ocean waves and eddies). While at McGill, I also received several équipe (team) awards and a Centre Grant from Québec's FCAR (Fonds pour la formation des Chercheurs et l'Aide à la Recherche). My students and I are very grateful for this substantial support during my years as a professor.

24.2 UBC Graduate Students

Daniel (Dan) G. Wright

Dan was the first in his family to attend university. He had a compact physique and described himself as a "military brat" who had to stand up for himself growing up on various Canadian Armed Forces bases. Professor Len Todd, his undergraduate mentor at Laurentian University in Sudbury, Ontario, advised him that if he wanted a challenge, he should do a PhD in applied mathematics at UBC under my supervision. Thus, Dan came to Vancouver in September 1975 as an NRC scholarship holder. He sailed through his graduate courses and PhD qualifying exam, while also supporting a wife and child. As a research assistant in the Mathematics Department, he meticulously checked all the derivations in the LeBlond-Mysak book *Waves in the Ocean* (Chap. 13). While doing this he formulated his own PhD thesis project on the dynamics of the Antarctic Circumpolar Current; he later published his results as the sole author. Remarkably, he completed his PhD in three years, during which time he and his wife had a second child.

I told Dan that he would make an excellent professor. He was creative, worked quickly, and explained problems very well. Dan also gave excellent seminars. "Nah, I prefer to do research," he replied. Indeed, he excelled as a research scientist for many years at the Bedford Institute of Oceanography in Halifax, Nova Scotia. When he came on sabbatical to McGill in 1989, he developed an elegant global ocean thermohaline circulation (THC) model which was the foundation of an ocean–atmosphere climate model widely used in paleoclimate studies [1] (see also Thomas Stocker below).

Dan was also instrumental in the formulation of "The International Thermodynamic Equation of Seawater-2010" (TEOS-10) that is now the standard for determining ocean water properties in modern oceanographic research. Regretfully, Dan died early in 2010, at age 58.

William W. Hsieh

William came to Vancouver from Hong Kong in the early 1970s to attend high school and university. As a senior at UBC, he caught my attention as an outstanding student in an advanced applied analysis course that I was teaching in 1976. "I want to become a theoretical physicist," he told me one day after class. After completing an MSc in this field under the support of an NSERC scholarship, he came knocking on my door in 1978 and asked about job prospects in oceanography. Positions were scarce in theoretical physics. Given his strong mathematics skills as well as his physics knowledge, I suggested that he do a PhD in physical oceanography focusing on weakly nonlinear wave–wave interactions in the ocean, a field which fascinated me at the time. This led to one joint publication [2] and two more papers by William on the analysis of oceanic data which provided evidence that such interactions between continental shelf waves occur in the ocean.

After completing his PhD in 1981, William did an original time series analysis of oceanic and fisheries data with Tim Parsons and me [3]. This paper helped pivot me into climate research. After postdoctoral work in England and Australia, William returned to UBC in 1986 as an assistant professor of oceanography and played a key role in the climate-fisheries project MOIST [4], which I started in 1984 (Chap. 15). Today, he is a professor emeritus at UBC and the proud author of two outstanding books on machine learning and environmental data analysis [5, 6]. Over the years, our families have enjoyed many wonderful Chinese dinners together.

Gordon E. Swaters

Gordon spent two years working on sea ice modeling at the Centre for Cold Oceans Resources Engineering located on the campus of Memorial University (St. John's, NFLD) before coming to UBC in 1981 for graduate studies in applied mathematics. Having worked in the field of oceanography, he knew about the then recent (1978) book *Waves in the Ocean* by LeBlond and myself and was keen to do research on waves and geophysical fluid dynamics under my supervision. Shortly after we met, I suggested that for an MSc thesis he should try to explain why the large (100-km diameter) Sitka Eddy west of the southern end of the Alaska Panhandle occurs as a semi-permanent feature of ocean circulation in the northeast Pacific Ocean. As the son of hard-working Dutch immigrants, Gordon loved challenges and proceeded to solve this problem by modeling the interaction of a horizontally sheared baroclinic current with a variable bottom topography consisting of an offshore seamount and a seaward protrusion of the continental slope [7]. Since I was on sabbatical in Zürich during 1982–83 (Chap. 14), Gordon worked independently on his thesis while I was away. In 1983, he moved on to a PhD program in applied mathematics. For his thesis he worked on the theory of "modons," a pair of dynamically coupled ocean eddies. He came up with this modon topic himself and only needed me as a sounding board; he finished his PhD in two years.

After a postdoc at MIT (1985–86), Gordon joined the mathematics department at the University of Alberta, my alma mater. He received early tenure in 1990 and attained full professor rank in 1993. Today he is internationally recognized for his fundamental contributions to our understanding of the dynamics of abyssal ocean currents, which are crucial for transporting heat, salt, and nutrients around the globe.

Andrew J. Weaver

Andrew, the eldest son of a British father and Ukrainian mother, was born in Victoria, BC just after I finished a summer job at the Pacific Naval Lab under the supervision of his father, John Weaver (Chap. 3). After completing his math-physics degree at UVic, Andrew applied to both the University of Cambridge and UBC for his PhD. His father was keen on Andrew going to Cambridge, a top British science university. He was in Cambridge for just one year (1983–84) and received a Master of Advanced Study degree in mathematics. "I really disliked the sherry and high tea socials at Cambridge," he once told me. He subsequently came to UBC to do an applied math PhD degree under my supervision, carrying out research on oceanic and atmospheric dynamics [8]. Andrew was an academically excellent and very enthusiastic student; he completed his PhD in three years. As with William Hsieh, Andrew was also a helpful team member of Project MOIST [4]. After postdocs in Australia (1987–88) and the United States (1989), followed by three years on the faculty at McGill as an NSERC University Research Fellow, Andrew returned to his home city as a professor of climate science at UVic in 1992. He quickly established himself as a leader in climate research and developed the widely used UVic Earth System Climate Model [9].

But Andrew wanted to do more for society than just science. In 2013, he was elected a Member of the Legislative Assembly (MLA) in BC as the first provincial Green Party representative in Canada, and during 2015–20, he was the party's leader. In the 2017 provincial election, he led the party to an historic result with 17% of the popular vote and three elected MLAs that held the balance of power in the provincial government. Consequently, many important environmental laws were adopted by the provincial government. After accomplishing his goals of developing a climate policy framework in BC (Clean BC) coupled with accountability legislation, he decided not to seek reelection and returned to UVic as a professor in 2020, where he is now concentrating on climate policy issues.

24.3 McGill Graduate Students

Davinder (Dav) K. Manak

Dav, a Canadian whose family came from India, was my first graduate student at McGill as well as my first *female* graduate student. After obtaining her honors BSc degree in physics from UVic, Davinder came to McGill in 1986 with the support of an NSERC scholarship on the recommendation of Andrew Weaver and his younger brother, Anthony, who were both doing research in oceanography. For her MSc thesis, Davinder did research on the interannual variability of the sea ice extent in the Arctic, which was a new area of study for me. Our first publication [10] has been widely cited because it established that during 1953–1984, the sea ice cover in each regional sea in the Arctic Ocean had a unique time scale of variability. Rather than going on for a PhD, Dav was happy to work at McGill as my research assistant (RA) for two years (1988–90) and be part of my growing group of lively graduate and postdoctoral students. She was a co-author of two notable papers on decadal climate variability in the Arctic [11] and on sea ice modeling [12]. In 1990 she moved to Ottawa to work in satellite remote sensing. Dav was employed by various federal government departments including the Canada Centre for Remote Sensing (CCRS), the Canadian Space Agency, and Natural Resources Canada. She co-authored numerous publications with other scientists at these institutions.

In 2023, I got a surprise email from Dav, who had just retired from the federal government. "I've recently been reminiscing about my McGill days and how it impacted my life and career," she wrote. "You were a tremendous support during my MSc and instrumental in landing my first job at CCRS. Thank you for being there during my time at McGill and your words of encouragement."

Silvia A. Venegas

In the winter of 1993, Ann Cossette, the secretary of the McGill climate centre, ran into Silvia Venegas wandering along the corridors of Burnside Hall, where my office is located. "Does anyone here work on El Niño?" Silvia inquired. "I am thinking of doing an MSc in climate, and I would like to work on a project related to this topic."

Silvia was a graduate in meteorology from Argentina who had been working as a weather forecaster in Montreal. She wanted to improve her qualifications and thus approached the recently formed C^2GCR regarding graduate studies. In September 1993, Silvia enrolled for an MSc in the AOS (Atmospheric and Oceanic Sciences) department. In the following year, David Straub (a new faculty member in AOS and member of C^2GCR) and I started jointly supervising her thesis on the interannual variability of ocean circulation off the coast of Argentina and its possible link to El Niño warming events. After completing her MSc in 1995, Sylvia was keen to apply the latest methods of analyzing atmospheric and oceanic climate data, which involved empirical orthogonal function and singular value decomposition analyses. Working as an RA with David and me, in consultation with the distinguished American climate scientist Michael Mann (the father of the "hockey stick" temperature record), she used the above methods to uncover the dominant patterns of *coupled* interannual variability in sea surface temperature and sea level pressure in the South Atlantic. These results were published in a paper [13] that has been cited over 400 times!

During her time as an RA, Silvia, who had a great artistic talent for preparing excellent colorful visuals, also helped me with my sea ice research [14, 15]. In 1998, she moved with her Danish husband to Copenhagen where she completed a PhD in Arctic climate dynamics. In 2006, they relocated to Argentina where Silva left science and used her artistic skills to design modern furniture. In 2021, the family returned to Denmark. "I like to change countries every now and then," she wrote to me recently. My wife Janet and I plan to visit Silvia and her family in Copenhagen in 2026.

L. Bruno Tremblay

After graduating from mechanical engineering at McGill in 1987, Bruno, a native of Chicoutimi, Quebec, spent a year in Switzerland as an exchange student. While studying at ETHZ, he did a project on ice sheet modeling with Koli Hutter, who was my host when I spent a sabbatical in Zürich in 1982–83 (Chap. 14). Koli knew that I had moved to McGill in 1986 to start a climate research program and told Bruno that I was looking for new graduate students. In 1992, Bruno took this to heart and came to talk to me about PhD studies in my department. His professional career had started as an aeronautical engineer at Pratt and Whitney in 1989, but he was now interested in pivoting into climate research if he could work on ice dynamics. "No problem," I said, "we need some new ideas in sea ice modeling, so welcome to McGill." Using his background knowledge in the science of materials, Bruno developed a novel dynamic sea ice model based on granular material rheology. He used this model to simulate well the commonly observed lead complexes (rectilinear cracks) in the sea ice cover of the Arctic Ocean [16]. The model was an improvement over earlier sea ice models and was later used in global climate models.

After completing his PhD in 1996, Bruno spent a decade at the Lamont Doherty Earth Observatory of Columbia University in New York working as a research scientist and lecturer in climate science. He kept in close contact with McGill and me during these years and helped supervise two of my MSc students working on the modeling of sea ice in the Arctic. In 2006, he returned to McGill as an assistant professor in AOS, specializing in Earth system science. Two decades on from 2006, Bruno is now a full professor supervising a stellar group of graduate students in several areas of Arctic climate and ocean research.

Zhaomin Wang

After eight years as a lecturer in atmospheric science at Nanjing University in China, Zhaomin came to McGill as a mature student in 1995 to do a PhD in the AOS Department. Inspired by the recent work in my group on climate and paleoclimate dynamics, Zhaomin developed in his PhD thesis a coupled ocean–atmosphere-sea ice-land surface model [17]. With the addition of an ice sheet model, he created the MPM, the McGill Paleoclimate Model [18], which was subsequently used by several of my graduate students who were co-supervised by Zhaomin. After completing his PhD in 1999, Zhaomin stayed on at McGill, first as my postdoc (1999–2001) and then as a Research Associate (see next section on Postdocs at UBC and McGill). The development of the MPM also resulted in important collaborations with climate scientists in Europe and North America who were research leaders in climate dynamics using similar types of models, now called Earth system Models of Intermediate Complexity (EMICs) [19].

Alexandra (Alex) Jahn

After completing her Diploma (MSc) at the Free University of Berlin in 2004, Alex applied to McGill for PhD studies in climate on the recommendation of her advisor Martin Claussen, from the Potsdam Institute of Climate Impact Research. When recruiting her to McGill, we had several calls on her "handy" (German expression for cell phone), as she wanted to make sure that she would be comfortable working with me as a supervisor. I knew she was also interested in studying at a couple of American universities, but I'm glad she came to McGill, where the plan was to use the MPM [18] to work on a project in paleoclimate dynamics. Alex was a hard worker and loved to explore new areas of research; but when she learned that Zhaomin was leaving McGill and would not be available to explain all the intricate details of the MPM, we reoriented her research toward studying the variability of the Arctic Ocean freshwater export with another global model, the UVic Earth System Climate Model (ESCM) [9]. As mentioned above, Bruno Tremblay joined the AOS faculty in 2006, and he became a co-advisor of Alex's PhD thesis. This was a fortunate development as he, together with Mike Eby at UVic, had the know-how to import a high-resolution version of the UVic ESCM to McGill. This collaboration led to the first publication from Alex's thesis [20]. Alex continued her study of freshwater exports from the Arctic using NCAR's Community Climate System Model Version 3 (CCSM3) with tracers now added to the model [21].

With her experience using the CCSM3, Alex moved to Boulder, Colorado, first as a postdoc and then as a project scientist at NCAR. In 2015 she joined the faculty of the University of Colorado in Boulder. She is now a tenured associate professor and leads an active group in polar and paleoclimate modeling. Alex is excellent at networking and an enthusiastic seminar speaker. As a PhD student and postdoc, she was actively involved in a polar interest group of young investigators (APECS—Association of Polar Early Career Scientists) that regularly met at international conferences.

Juliana M. Marson

After I retired from teaching at McGill in 2010 and was winding down my research group, I got a surprise email from Juliana in spring 2013. "I have funding for a year in my program of PhD studies at the Universidade de São Paulo (Brazil) to be a visiting graduate student anywhere in the world," she wrote. "Would you be able to provide an office, a computer and some guidance on a project in paleoclimate dynamics?"

How could I refuse? So, I took a chance and invited her to join my now reduced research group at McGill. Juliana turned out to be a wonderful visiting graduate student. She came into the office daily, worked with a smile and enjoyed going to a couple of Canadian conferences where she started networking. While she was at McGill for the 2013–14 academic year, she applied an interesting statistical mixing triangle method to analyze the evolution of the deep-water masses in a 21-Kyr run of the NCAR CCSM3. The goal was to study how the simulated North Atlantic Deep Water and the Antarctic Bottom Water changed during (i) the first cold Heinrich event in the North Atlantic (from 19 to 14 Kyr ago), (ii) the following abrupt warm Bølling-Allerød period, and then (iii) the sudden cold period known as the Younger Dryas (12.9 to 11.7 Kyr ago). An important service I provided to Juliana while at McGill was teaching her how to navigate the refereeing process for publishing a paper in a top climate journal [22]. We also enjoyed talking about the lives of other scientists and how to interact with our colleagues.

After finishing her PhD in 2015 in Brazil, Juliana wanted to return to Canada. I provided a strong reference for her to work as a postdoc at the University of Alberta with Paul Myers, one of my academic grandchildren (Paul was a PhD student of Andrew Weaver). She later performed very well at the U of A as a sessional lecturer in oceanography and climate. In 2021, Juliana secured a tenure-track position at the University of Manitoba, where she now leads an active research group focusing on the polar oceans and their interactions with the cryosphere and climate. She has also become an expert in iceberg motion. Juliana and her Brazilian husband now have two young children and are very busy parents.

24.4 Postdoctoral Students at UBC and McGill

Andrew J. Willmott

Andrew came to UBC in 1979 on the recommendation of his PhD supervisor Professor John Johnson from the University of East Anglia (Norwich, UK), whom I had met when I spent a sabbatical in Cambridge during 1971–72 (Chap. 11). There was some drama upon his arrival in Vancouver. I had arranged short-term accommodation for Andrew and his fiancée Sasi at the home of a friend who was away. Unfortunately, the friend had locked the front door with a dead bolt, and the key he gave me was now useless. However, we discovered a partly open basement window at the back of the house, and Andrew was able to force his way in, breaking windowpanes in the process. Luckily, no one heard the ruckus during this "break-in entry," and we were able to repair the window before the friend returned home.

Andrew was a very enthusiastic and hard-working collaborator who was keen to investigate the eddy-like motions in the northeast Pacific Ocean [23] and the generation of trench waves [24]. He and Sasi became good friends of the family, and they often babysat our young children, Paul and Claire. In 1981, Andrew obtained a tenure-track position at the prestigious Naval Postgraduate School in Monterey, CA where he began working on Arctic Ocean dynamics. They returned to England a few years later, and over the next few decades, Andrew held senior academic positions at Exeter and Keele universities and was later Director of Science of the Natural Environmental Research Council Proudman Laboratory in Liverpool. During his tenure there, the laboratory became the National Oceanographic Centre (NOC), with sites in Liverpool and Southampton, and Andrew became Director of Science and Technology for NOC. He retired in 2023 as Head of the School of Mathematics, Statistics, and Physics at the University of Newcastle. We have often visited each other over the years and still exchange Christmas cards.

Moto Ikeda

Moto came to UBC in 1981 in response to an advert Bill Emery and I posted looking for a postdoc who had experience in both satellite remote sensing of the oceans and ocean modeling. Moto had a PhD in aeronautical engineering from Japan and had moved to the US in the late 1970s to begin research in remote sensing. He arrived in Vancouver with his wife Haruhi, two young sons, and a large traditionally dressed Japanese doll in a glass cage, which was a gift to me and my then wife, Mary. My daughter Claire now has this beautiful doll in her home.

Moto, Bill, and I investigated the meanders and stability of the California Current System off Vancouver Island [25]. Moto was a meticulous worker who was always focused on the details of both the data and the model results. After two years at UBC, he moved to the Bedford Institute of Oceanography in Halifax, NS where he spent over a decade (1983–95) working as a research scientist on the dynamics of the Arctic and the North Atlantic oceans. He returned to Japan in 1995 to lead a new faculty of earth and ocean sciences at Hokkaido University in Sapporo (on the island of Hokkaido, Japan). Through his contacts in North America, he invited many visitors to Sapporo, including me (as a visiting professor in spring 1997 under the sponsorship of the Japan Society for the Promotion of Science). Moto was a gracious host in Sapporo, and he also arranged for me to give lectures in five other ocean research centres in Japan. We have kept in contact over the years, and we met regularly at annual SCOR meetings when I was president of IAPSO (2007–11) (Chap. 22).

Thomas F. Stocker

Thomas came to McGill in 1989 on the recommendation of my Swiss colleague Koli Hutter. At ETHZ during the mid-1980s, Koli supervised Thomas's outstanding PhD thesis on topographic waves; Koli suggested that climate change would be a good topic for postdoctoral studies. Thomas took up the challenge and was successful in obtaining a prestigious Postdoctoral Fellowship from the Swiss National Science Foundation, allowing him to spend two years at McGill (1989–91). While I had investigated natural climate variability on the interannual to decadal timescales, I knew little about natural climate variability on the century timescale. Since climate change due to the burning of fossil fuels and land-use change was occurring on the latter timescale, I thought it was important to look at long proxy-climate data records from tree rings and ice cores to see whether natural climate variability also occurred on this timescale. Studying paleoclimate data records was a learning process for both of us, and in the end, we provided evidence of this type of variability [26]. To advance the modeling of this variability, we were fortunate that Dan Wright, one of my brilliant PhD students at UBC (see above), chose to spend a sabbatical at McGill while Stocker was here. With his deep knowledge of dynamic oceanography, Dan formulated a new zonally averaged ocean model of the global thermohaline circulation (THC) (also known as the conveyor belt) which Thomas coupled to an atmospheric energy balance model [1]. This was the birth of global climate models of intermediate complexity [19]. The model [1] was then used, for example, to study changes in the THC caused by large freshwater discharges from Greenland Icesheet, and more generally, to understand long-term paleoclimate changes.

In 1991, Thomas moved to the Lamont Doherty Earth Observatory in New York to work with Wally Broecker on the global carbon cycle and paleoclimate dynamics. In 1993, at age 34, he became professor of climate and environmental physics at the University of Bern, a position formerly held by the distinguished polar scientist Hans Oeschger who invented a high-precision beta-decay counter to measure rare radionuclides in the environment, such as 14C and 39Ar. In Bern, Thomas led a very active and prolific research group in climate dynamics, paleoclimates, and climate reconstructions until his retirement in 2024. Stocker is also widely recognized for his outstanding leadership in editing the physical scientific basis for the Fifth (2013) Assessment Report of IPCC [27].

Over the three decades after his time at McGill, Thomas has kept closely in touch. In September 2023, Thomas and his wife Letizia hosted Janet and me at their beautiful home in Bern prior to our cycling adventure around Bodensee (Lake Constance), a large lake which lies between Switzerland, Austria, and Germany.

Gavin A. Schmidt

Like Thomas Stocker, Gavin also did his PhD on topographic waves, but at University College London under the supervision of Professor Ted Johnson, who was my postdoc at UBC during the late 1970s. In late December 1993, when it was − 10 °C outside, Gavin arrived at our Montreal West home directly from the airport dressed in a Harris tweed sports jacket. "I think I'll be needing a warmer winter coat," he said as he entered our home. Gavin was a quick learner!

Gavin and I had corresponded before his arrival at McGill about possible research topics. I was keen on learning more about the natural variability of the THC and its application to paleoclimates. Gavin, having done his PhD in a mathematics department, confessed that he had never heard of the THC. However, he soon became very knowledgeable on the structure and properties of the zonally averaged THC model developed by Dan Wright. We investigated the dynamic stability of this model [28] and applied it to the land–ocean configuration of the Cretaceous period 100 million years ago, when there was no Atlantic Ocean, only a very wide proto–Pacific Ocean [29]. We were able to show that there was a THC pattern that warmed the polar regions to an extent that is consistent with ancient ocean data collected from deep ocean sediment cores.

Gavin moved to NASA's Goddard Institute for Space Studies (GISS) in New York in 1996, first as a NOAA postdoctoral fellow and then as an Associate Research Scientist at nearby Columbia University. He became a NASA civil servant at GISS in 2004 and co-founded RealClimate blog, a highly appreciated public commentary site on climatology. He replaced the renowned James Hansen as Director of GISS in 2014. As well as doing highly cited research in Earth system model development, recent climate change, paleoclimates, and water isotopes in the climate system, Gavin excels in science media and outreach. In 2011 he was awarded the inaugural AGU Climate Communications Prize.

Zhaomin Wang

As mentioned earlier in the McGill graduate student section, Zhaomin stayed on at McGill as a postdoc (1999–2001) when we continued working with the MPM on various paleoclimate phenomena such as glacial inception [30], the cold climate

THC [31] and THC-ice sheet interactions [32]. As a Research Associate (2001–2005), Zhaomin was an outstanding co-supervisor of several of my graduate students who investigated vegetation dynamics in the MPM [33], ice sheet discharges [34] and long-term future climate changes in the MPM [35]. Before he left McGill, we co-authored one last joint paper on glacial abrupt climate change [36].

In 2006, Zhaomin moved to the British Antarctic Survey (BAS) in Cambridge, England as a Research Scientist where he developed expertise in modeling the Southern Ocean circulation and its connections to Antarctic ice dynamics. His experience at McGill was highly regarded by the BAS and played a pivotal role in his securing a permanent position there. In 2012, he returned to China as professor of ocean and climate science at the Nanjing University of Information and Technology where he took a leading role in advancing polar ocean research in China. Since 2022, he has served as professor and head of the polar ocean circulation group at the Southern Marine Science and Engineering Guangdong Laboratory in Zhuhai.

24.5 Epilogue

After finishing their graduate degrees or postdoc positions at UBC or McGill, the above students have been recognized with prizes, awards, and honors (both national and international) for their outstanding research. These are too numerous to list here. The same is true of many of my other students whom I have not included in this chapter. I apologize for not mentioning them here. But I want them to know that I still have many happy memories of the times when they were my students at UBC or McGill.

Acknowledgements I thank Janet Boeckh for her comments on various drafts of this chapter, and all the students who reviewed their profiles: in order of appearance, these include William Hsieh, Gordon Swaters, Andrew Weaver, Davinder Manak, Sylvia Venegas, Bruno Tremblay, Zhaomin Wang, Alexandra Jahn, Juliana Marson, Andrew Willmott, Moto Ikeda, Thomas Stocker, and Gavin Schmidt. I am also grateful for the feedback I received from Claudia Bierman, Roussos Dimitrakopoulos, Mary Feher, Djordje Romanic, Neil Stewart, and Richard Thomson.

References

1. Stocker TF, Wright DG, Mysak LA (1992) A zonally averaged, coupled ocean-atmosphere model for paleoclimate studies. J Climate 5:773–797
2. Hsieh WW, Mysak LA (1980) Resonant interactions between shelf waves, with application to the Oregon shelf. J Phys Oceanogr 10:1729–1741
3. Mysak LA, Hsieh WW, Parsons TR (1982) On the relationship between interannual baroclinic waves and fish populations in the Northeast Pacific. Biol Oceanogr 2:63–103
4. Mysak LA, Groot C, Hamilton K (1986) A study of climate and fisheries: interannual variability in the northeast Pacific Ocean and its influence on homing migration of sockeye salmon. Climatological Bull 20:26–35

5. Hsieh WW (2009) Machine learning methods in the environmental sciences: neural networks and Kernels. Cambridge University Press, Cambridge
6. Hsieh WW (2023) Introduction to environmental data science. Cambridge University Press, Cambridge
7. Swaters GE, Mysak LA (1985) Topographically induced baroclinic eddies near a coastline, with application to the northeast Pacific. J Phys Oceanogr 15:1470–1485
8. Weaver AJ, Mysak LA, Bennett AF (1988) The steady state response of the atmosphere to midlatitude heating with various zonal structures. Geophys Astrophys Fluid Dyn 41:1–44
9. Weaver AJ, Eby M, Wiebe EC, Bitz CM, Duffy PB, Ewen TL, Fanning AF, Holland MM, MacFadyen A, Matthews HD, Meissner KJ, Saenko O, Schmittner A, Wang H, Yoshimori M (2001) The UVic Earth system climate model: model description, climatology, and applications to past, present and future climates. Atmos-Ocean 39:361–428
10. Mysak LA, Manak DK (1989) Arctic sea-ice extent and anomalics 1953–84. Atmos Ocean 27:376–405
11. Mysak LA, Manak DK, Marsden RF (1990) Sea-ice anomalies observed in the Greenland and Labrador seas during 1901–1984 and their relation to an interdecadal Arctic climate cycle. Clim Dyn 5:111–133
12. Holland DM, Mysak LA, Manak DK, Oberhuber JM (1993) Sensitivity study of a dynamic thermodynamic sea-ice model. J Geophys Res 98:2561–2586
13. Venegas SA, Mysak LA, Straub DN (1997) Atmosphere-ocean coupled variability in the South Atlantic. J Climate 10:2904–2920
14. Mysak LA, Venegas SA (1998) Decadal climate oscillations in the Arctic: a new feedback loop for atmosphere-ice-ocean interactions. Geophys Res Lett 25:3607–3610
15. Venagas SA, Mysak LA (2000) Is there a dominant timescale of natural climate variability in the Arctic? J Climate 13:3412–3434
16. Tremblay L-B, Mysak LA (1997) Modeling sea ice as a granular material, including the dilatancy effect. J Phys Oceanogr 27:2342–2360
17. Wang Z, Mysak LA (2000) A simple coupled atmosphere-ocean-sea ice-land surface model for climate and paleoclimate studies. J Climate 13:1150–1172
18. Mysak LA, Wang Z (2000) The McGill Paleoclimate Model (MPM): a new Earth system model of intermediate complexity. CMOS Bulletin SCMO 28:104–109
19. Claussen M, Mysak LA, Weaver AJ, Crucifix M, Fichefet T, Loutre M-F, Weber SL, Alcamo J, Alexeev VA, Berger A, Calov R, Ganopolski A, Goose H, Lohmann G, Lunkeit F, Mokhov II, Petoukhov V, Stone P, Wang Z (2002) Earth system models of intermediate complexity: closing the gap in the spectrum of climate system models. Clim Dyn 18:579–586
20. Jahn A, Tremblay B, Mysak LA, Newton R (2010) Effect of the large-scale atmospheric circulation on the variability of the Arctic Ocean freshwater export. Clim Dyn 34:201–222. https://doi.org/10.1007/s00382-009-0558-z
21. Jahn A, Tremblay B, Newton R, Holland MM, Mysak LA, Dmitrenko IA (2010) A tracer study of the Arctic Ocean's liquid freshwater export variability. J Geophys Res 115:C07015. https://doi.org/10.1029/2009JC005873
22. Marson JM, Mysak LA, Mata MM, Wainer I (2016) Evolution of the deep Atlantic water masses since the last glacial maximum based on a transient run of NCAR-CCSM3. Clim Dyn 47:865–877. https://doi.org/10.1007/s00382-015-2876-7
23. Willmott AJ, Mysak LA (1980) Atmospherically forced eddies in the Northeast Pacific. J Phys Oceanogr 10:1769–1791
24. Mysak LA, Willmott AJ (1981) Forced trench waves. J Phys Oceanogr 11:1481–1502
25. Ikeda M, Mysak LA, Emery WJ (1984) Observation and modelling of satellite-sensed meanders and eddies off Vancouver Island. J Phys Oceanogr 14:3–21
26. Stocker TF, Mysak LA (1992) Climatic fluctuations on the century time scale: a review of high-resolution proxy data and possible mechanisms. Clim Change 20:227–250
27. IPCC (2013) Climate Change 2013: The physical science basis. In: Stocker TF, Qin D et al (eds) Contributions of the working group I to the fifth assessment report of the intergovernmental panel on climate change. Cambridge University Press, Cambridge, UK and New York, NY, USA, pp 1535 (First published in 2014)

28. Schmidt GA, Mysak LA (1996) The stability of a zonally averaged thermohaline circulation model. Tellus 48A:158–178
29. Schmidt GA, Mysak LA (1996) Can increased poleward oceanic heat flux explain the warm Cretaceous climate? Paleoceanography 11:579–593
30. Wang Z, Mysak LA (2002) Simulation of the last glacial inception and rapid ice sheet growth in the McGill Paleoclimate Model. Geophys Res Lett 29(23):2102. https://doi.org/10.1029/2002GL015120
31. Wang Z, Mysak LA, McManus JF (2002) Response of the thermohaline circulation to cold climates. Paleoceanography 17(1):6. https://doi.org/10.1029/2000PA000587
32. Wang Z, Mysak LA (2001) Ice sheet-thermohaline circulation interactions in a climate model of intermediated complexity. J Oceanogr 57:481–494
33. Wang Y, Mysak LA, Wang Z, Brovkin V (2005) The greening of the McGill Paleoclimate Model. Part I: improved land surface scheme with vegetation dynamics. Clim Dyn 24:469–480. https://doi.org/10.1007/s00382-004-0515-9
34. Papa BD, Mysak LA, Wang Z (2006) Intermittent ice sheet discharge events in northeastern North America during the last glacial period. Clim Dyn 26:201–216. https://doi.org/10.1007/s00382-005-0078-4
35. Cochelin A-SB, Mysak LA, Wang Z (2006) Simulation of long-term future climate changes with the Green McGill paleoclimate model: the next glacial inception. Clim Change 79:381
36. Wang Z, Mysak LA (2006) Glacial abrupt climate changes and Dansgaard-Oeschger Oscillations in coupled climate model. Paleoceanography 21:PA2001. https://doi.org/10.1029/2005PA001238

Chapter 25
Chairing the Steacie Prize Selection Panel

Abstract Since 2013, I have had the honor of serving as Chair of the Steacie Prize Selection Panel, which each year recognizes a scientist or engineer aged 40 or younger for outstanding research done in Canada. At a public ceremony held in their home institution, the recipient of the Prize receives an award certificate, a cheque, and a copy of the biography of Steacie by M.C. King. The Prize was established in 1963 in honor of E.W.R. Steacie FRS FRSC, an eminent Canadian chemist, who served with distinction as president of the National Research Council of Canada during 1952–62. The Prize is administered by the Trustees of the Steacie Memorial Fund, which is a a private charitable foundation. The quality of the past winners of the Prize is exceptional—nearly all become FRSC and about 25% become FRS. For these reasons, the Prize is widely considered as Canada's most prestigious early career award for scientists and engineers.

25.1 Introduction

At 08:46 on 6 May 2025, the VIA Rail train starts rumbling out of the dark underground cavern of Montreal Central Station. I am travelling to Toronto with my wife Janet to present the 2024 Steacie Prize to Dr. Jo Bovy, a brilliant astrophysicist at the University of Toronto. Named after the distinguished Canadian chemist Edgar WR Steacie, the prize is awarded annually to a scientist or engineer aged 40 or younger for outstanding research done in Canada. Many colleagues regard it as Canada's most prestigious early career award. As chair of the Steacie Prize Selection Panel, I have the honour of presenting the prize to the winner at a ceremony at the recipient's home institution. Janet and I are very excited about this trip; we have never met Dr. Bovy and we're looking forward to the prize ceremony and his public lecture tomorrow on the evolution and structure of the Milky Way.

L. Mysak, *Adventures in Climate Science, Ocean Waves, and the Flute*, Springer Biographies, https://doi.org/10.1007/978-3-032-19848-8_25

25.2 Joining the Selection Panel

In July 2010, I got a call from Bob McKellar whom I had worked with in the mid 1990s when I was president of the Academy of Science, RSC. Bob was then the secretary for the Steacie Memorial Fund. He asked me to serve on the Steacie Prize Selection Panel because of my experience in serving on or chairing several award and prize committees. I had just retired from McGill and would have time available. He told me that the chair of the selection panel was ecologist John Smol, whom I had inducted into the RSC in the mid 1990s. A small world indeed! I agreed to serve on the panel for three years, and since 2013, I have been its chair.

25.3 Edgar (Ned) Steacie

Steacie was instrumental in promoting Canadian science during the middle of the 1900s. He was born in Montreal on Christmas Day, 1900. He completed a PhD in physical chemistry at McGill in 1926 and joined its chemistry department in 1928. To quote science historian Christine King [1], "Steacie swiftly rose in the world of Canadian science." In 1939 he moved to Ottawa to become Director of the Chemistry Division of the National Research Council (NRC). During World War II, he was second in command to Sir John Cockcroft at the British Canadian Atomic Energy Project in Montreal.

After the war, Steacie's NRC lab became famous for the study of chemical kinetics. In 1948 he launched the NRC postdoctoral fellowship program, which attracted 14 young scientists from around the world, many of whom later became Fellows of the RSC. In 1952 he became President of the NRC (Photo 25.1), serving with distinction in this position until his death in 1962.

During Steacie's presidency, the rise in Canada's position as a scientific nation was phenomenal [1]. Thanks to his leadership, the laboratories of NRC in Ottawa became renowned centres of excellence. He believed that Canada should promote and expand its own university graduate programs in science and engineering, and accordingly, he introduced graduate and postdoctoral fellowships tenable in the universities. In addition, NRC awarded its first "Operating Grants" to university professors to support basic research in science and engineering. Our current national funding agency NSERC was spun off from NRC in the late 1970s. Today, the Operating Grants are called "Discovery Grants," and NSERC continues to award graduate and postdoctoral fellowships. For five decades I and my research group have benefited from these grants and fellowships.

After Steacie's untimely death in 1962, the prominence of science in Canada has followed an up and down pattern, and the value of science to society has often been questioned. Fortunately, the present Liberal federal government has generally shown a healthy regard for science, which sadly is not the situation in the USA today. Examples of the Canadian government's respect for science include the appointment

Photo 25.1 Edgar (Ned) Steacie FRS FRSC as the new President of the National Research Council of Canada. This picture is from the book cover of King's biography of Steacie [1]. Permission granted from University of Toronto Press

of a Chief Science Advisor, the establishment of Collaboration Centres between NRC and university scientists, and the application of science-based policies to deal with the COVID-19 pandemic during 2020–22. As a follow-up of this pandemic, two of our national academies, the Royal Society of Canada, and the Canadian Academy of Health Sciences, put together an expert panel in 2023 to review Canada's past, present, and future role in advancing health globally. On 27 March 2025 they released their impressive report [2].

25.4 The Steacie Prize and Its Administration

The Steacie Prize was established in 1963 through the financial contributions of several hundred friends, associates and former colleagues of Steacie. The cash component of the Prize comes from the income of the EWR Steacie Memorial Fund, a private charitable foundation which is administered by the Trustees of the Fund, who are listed on the website for the Steacie Prize (https://www.steacieprize.ca). The foundation is dedicated to the advancement of science and engineering in Canada.

Each year a selection panel chair and six distinguished scientists and engineers from across the country are appointed to choose the winner. Three are usually in the life sciences domain and three in the physical sciences. Most members of the selection panel are Fellows of the RSC, and often there are two or three former Steacie Prize winners on the panel.

25.5 Past Winners of the Prize

The first Steacie Prize was awarded to University of Toronto physicist Jan Van Kranendonk in 1964, and in the following year it went to John Polanyi, also from U of T. Polanyi subsequently won the Nobel Prize for Chemistry in 1986. The complete list of previous winners can be found on the Steacie Prize website (see above). Verena Tunnicliffe, a deep-sea oceanographer from the University of Victoria, was the first woman to win the prize in 1993. Since I have served on the selection panel, six women have received the prize. As of 2025, 66 prizes have been awarded, with over half (37) going to physicists, chemists, or mathematicians. Life scientists have received 22 of the prizes, and the remaining (7) have gone to engineers or environmental scientists. Over half of the prizes have been won by U of T or UBC professors, but scientists and engineers from NRC as well as from many other universities from coast to coast have also received the prize. The average age of the recipients is 38.

On a few occasions two prizes have been awarded in the same year, but the general policy of the selection panel is to choose only one winner. During the past decade, there have been 25 to 40 nominations of excellent candidates each year, and choosing just one winner is a big challenge. To reflect the high quality of the recent nominations, the panel may now name up to three finalists for each year's prize.

As an indication of the wide range of fields, and regional and gender distribution of the winners since 2019, I list below their names, fields of research, and home institutions (Photo 25.2):

2024 Jo Bovy, Astrophysics, University of Toronto.
2023 Asim Biswas, Soil Science, University of Guelph.
2022 Meghan Azad, Child Development, University of Manitoba.

Catherine Lebel, Radiology, University of Calgary.

2021 Erin Johnson, Theoretical Chemistry, Dalhousie University.
2020 Paul McNicholas, Statistics, McMaster University.
2019 Mary O'Connor, Marine Biodiversity, UBC.

Photo 25.2 The author presenting the 2022 Steacie Prize Certificate to Professor Catherine Lebel at the University of Calgary on 26 April 2023

25.6 The Selection Process

Nominations for the Steacie Prize, due September 1, are sent to the secretary of the Steacie Memorial Fund, currently Linda Johnston, who in late September sends each selection panel member an electronic file with the nomination letters, CVs, and reference letters for all the candidates. Because of the large number of nominations in recent years, the file is subdivided into a life sciences group and a physical sciences group which then are reviewed respectively by the three life science panel members and the three physical science panel members. At this stage the quality of the reference letters plays an important role in arriving at two short lists of candidates who have done innovative, impactful, and significant research. These two lists are next reduced into one list, and these candidates are then ranked and discussed by the whole panel at a Zoom call in late November, during which a winner is chosen. As mentioned above, up to three finalists may also be identified at this meeting.

After the Zoom call, I phone the winner, who is always thrilled to hear the good news. To quote from the news release 9 January 2025 for the 2024 winner Jo Bovy, "I am incredibly honoured to receive this prize as it is a great recognition of my work. As someone working in a field with few immediate practical applications, I personally find it rewarding to see that pure, curiosity-driven research into our cosmic origins and the fundamental nature of matter is appreciated by a wider community."

25.7 Award Ceremony

After I notify the prize recipient and the Steacie Fund secretary, I contact the winner's nominator to begin the process of organizing a public award ceremony where I (or another Trustee of the Memorial Fund) will present the prize certificate, King's biography [1], and a cheque. This ceremony is followed by the winner's public lecture. My wife Janet often comes with me to the award ceremony and lecture as she enjoys learning about the cutting-edge advances in science and engineering. After all, we first met at a Cutting-Edge Science Lecture Series at McGill in January 2012 (Chap. 21).

Our trip to Toronto in May 2025 followed the routine of earlier Steacie Prize presentation events. After our arrival at Union Station, we made our way by taxi to the boutique Annex Hotel, just north of the University of Toronto. That evening we treated ourselves to a performance of the Lion King, a lively and colorful musical. Next day, Janet and I separately visited old friends who live in Toronto; in my case I enjoyed a cappuccino with Henry Auster whom I met at Harvard in the 1960s when we were both graduate students there. He retired a few years ago as a professor of English at the U of T.

After lunch Janet and I headed for the Astronomy and Astrophysics Department in search of Jo Bovy. He was casually dressed in a long-sleeved top and jeans; in my suit and tie, I was clearly overdressed for the occasion (as Janet suspected I would be). After a friendly chat, we made our way to Innis College on the U of T campus where the award ceremony was to take place. After speeches by the department chair, the associate vice president research, the dean of science and me, I presented the Steacie Prize certificate, a copy of King's biography of Steacie [1], and a cheque for $10,000 to Dr. Bovy. We then moved into a large lecture hall where 200 guests were eagerly waiting to hear Bovy's dynamic and colorfully illustrated lecture on the origin, evolution and dynamics of the Milky Way. "What a brilliant lecture," Janet said afterward. "My head will be spinning for a long while." The lively Q&A session afterward lasted for over half an hour.

At around 7 pm, about ten of us (including Mrs. Bovy) walked to the Piano Piano Restaurant for a sumptuous Italian meal. It was a very noisy place, just the opposite of what piano piano means in music, namely pianissimo, or very quiet. The very tasty food courses were varied and kept coming all evening. By 10.30 pm we were the only ones left in the restaurant, and the sudden peace and quiet was much enjoyed. Shortly after, we dispersed; Janet and I had a relaxing 10-min walk to our hotel through the charming Annex district, which is home to many U of T professors, artists, and writers.

25.8 Epilogue

Looking back over the six-decade history of the prize, it is remarkable that nearly all winners become FRSCs and about 25% have been elected Fellows of the Royal Society of London (FRS). Thus, the quality of the past winners is exceptional, which is why the Steacie Prize is so prestigious. I like to call it Canada's "mini-Nobel prize" for the young.

In my view, it would be fitting if the Steacie Memorial Fund could provide a larger cash prize. For this to happen, more donations would have to be made to the foundation. As noted in the Steacie Prize website, these donations should be directed to the treasurer of the Memorial Fund. Since the foundation is a registered charitable organization, the treasurer will issue a tax receipt to any donor.

I have thoroughly enjoyed the 12 years that I have served as chair of the selection panel. Meeting the winners is always a highlight of the year for me and my wife. However, I do believe that it is important to bring in new blood into the position. While I can officially continue as chair until 2028, I expect that I will want to step down a year or two before then.

Acknowledgements I thank Janet Boeckh for reviewing the first draft of this chapter. The comments and feedback from Jacques Derome, Linda Johnston, Bob McKellar, Milica Radisic, John Smol, and Dan Wayner are gratefully appreciated.

References

1. King MC (1989) EWR steacie and science in Canada. University of Toronto Press, Toronto
2. Evans T, Lee K, Blouin C, Caron N, Clark J, Greenhill R, Liu J, Omaswa R, Philpott J, Reddy K (2025) Protecting our collective future: renewing Canada's role in global health. Royal Society of Canada and Canadian Academy of Health Sciences, Ottawa

Part III
Other Adventures and Loves

Chapter 26
Mastering Tone and Technique on the Flute

Abstract In the summer of 1958, when I was 18, I spent six weeks at the famed Aspen Music School in Colorado studying flute performance with the principal flutist of the Detroit Symphony. In those six weeks, I completely changed my tone to a richer and softer sound, and I received very useful advice on how to perform in public like a professional. These skills I have used throughout my musical life, enabling me to play flute in excellent amateur orchestras and give concerts in various venues. Even in my mid-eighties I am still able to play with confidence and a warm tone.

I started on low G, playing without a break, the G-Major scale over the three-octave range of the flute. Then I repeated the scale, but tonguing each note instead: tu, tu, tu, tu, etc. Finally, I played the scale using the articulation of slurring two notes and tonguing two: taha, tu, tu, taha, tu, tu, etc.

"Your technique is good, but your tone is harsh," commented Albert Tipton, principal flute with the Detroit Symphony. "Please give me your flute." Mr. Tipton was about 40 years old, athletic and very tall—well over six feet. He peered down at me, a scrawny 18-year-old teenager, and he yanked off the head joint. "For the next week," he said, "I want you *only* to practice blowing over the hole in the head joint. Try to develop a more mellow, relaxed sound."

Tipton then produced the tone that I should strive for. "At the next lesson we will check your tone, and if it's improved, we'll start to work on a simple Handel Sonata."

I was stunned. This was summer 1958, nearly seven decades ago, when I was studying advanced flute under Professor Tipton at the famed Aspen Music School in Colorado. Earlier in the spring of that year, I completed my final exam in Edmonton for the Performance Flute Diploma from the University of Alberta in Edmonton. For the exam I played the elegant but challenging G-Major Mozart Flute Concerto, the demanding Flute Sonata No. 1 by JC Bach, and the devilishly difficult Sonata for Flute Alone by the American composer Virgil Thomson. I got 85% on my final exam, a first-class honors grade. But not one of the examiners commented on the quality of my tone, which I had developed over the past six years. What a shock to be told by Mr. Tipton to go back to the drawing board to make it less harsh.

L. Mysak, *Adventures in Climate Science, Ocean Waves, and the Flute*, Springer Biographies, https://doi.org/10.1007/978-3-032-19848-8_26

That week was extremely frustrating. Ask yourself, "how many ways can you make a sound by blowing over the top of an open coke bottle?"

Many, as it turns out. Each day I positioned my lips over the opening of the head joint and blew. One day the lips were tight. Another day, they were loose. I tried blowing horizontally across the hole and then more into it. I rotated the head joint away from me, then toward me. I lost count of all the different positions I tried. Miraculously, on the fifth day, I produced a beautiful mellow sound. *This should please Mr. Tipton*, I thought.

Yes, at the second lesson Mr. Tipton was delighted with my improved tone. What a relief, since earlier that week I was ready to head back to Edmonton, as one very disappointed flute player.

Learning Handel's Flute Sonata No. 4 under Tipton was a joy. Because the music is not too difficult technically, I could concentrate on using my new tone when playing the lyrical melodies in the sonata. An accomplished pianist, Tipton played the piano accompaniment while I did the solo line. At the end of the summer, I wanted to give a recital to show off my new sound.

For public performance Mr. Tipton gave me excellent advice. "Never play a piece in which you have difficulty playing all the notes. It's much better to drop down a level, and play a simpler piece flawlessly," he said. The Handel sonata suited this philosophy perfectly.

I never did a recital at Aspen, but when I returned to Edmonton in the fall, I played the Handel sonata with ease and enjoyment at concerts with Lynne Newcombe, an outstanding piano student at the University of Alberta who was in the inaugural year of the Bachelor of Music program. We were warmly applauded at our performances. Lynne and I also made a vinyl recording of the sonata, which I gave to my parents as a present in 1959. When my 96-year-old dad moved into a senior's residence around 2002, the record was rescued by my sister, Helen. A few years later, she passed it on to her daughter, Daphne. In 2018, I made a digital copy of this ancient recording at the music school of McGill University.

Although I benefitted enormously from my Aspen experience, I decided that a career as a flutist was not for me. There is just too much talent out there to compete with. But I put my new tone to good use over the next seven decades. I have played in chamber groups, entertained seniors in their residences, and given solo performances at birthdays. Also, I was fortunate to have played flute for three decades in the I Medici di McGill Orchestra (Chap. 27), a high-level amateur orchestra which is now led by the internationally recognized Maestra Lori Antounian. The previous conductors of this orchestra include Wanda Kaluzny, Iwan Edwards, Gilles Auger and Veronique Lussier.

That summer back in Aspen, I do remember very well playing in the student orchestra under one conducting student. He was about 15 in 1958, and he ripped us through Haydn's London Symphony No. 104 at neck-breaking speed. *I wondered where he would end up.* It turned out to be the New York Metropolitan Opera. Yes, in summer of 1958, I played under Maestro James Levine who served as the Met's Music Director for 40 years, from 1976 to 2016.

26.1 Coda

This chapter is an update of what I wrote as an assignment for the Thomas More Institute (TMI) Memoir Writing Course that I took in winter 2018. I am very grateful to Pauline Beauchamp and Karen Nesbitt, the course instructors, for their feedback and suggestions for the improvement of this short memoir.

Chapter 27
A Call from the Orchestra Director

Abstract In September 1994, I was invited to join the flute section of the I Medici di McGill Orchestra, a "doctors" amateur symphony founded in 1989 by McGill pharmacology professor Ante Padjen. For 30 years, I have had much pleasure playing a broad range of symphonic repertoire and accompanying many outstanding soloists. The orchestra is based in McGill's Faculty of Medicine because in 1989, Dr. Padjen proposed to Richard Creuss, the then dean of the faculty, that an orchestra would offer Montreal audiences the healing power of music through concerts benefiting the healthcare community and various charities. Some of the organizations and charities that have benefitted from I Medici concerts include the Montreal Association for the Blind, the Mackay Foundation (for disabled children), Syrian refugees, the Médicins Sans Frontières in Sudan, the Bell Fund for Cancer, and the Cedars Cancer Foundation. The orchestra has also given two concerts for the medical benefit of people in Ukraine affected by the brutal invasion by Russia in February 2022.

I officially retired from the orchestra in May 2024 after having played under the first four conductors of I Medici. However, I did return as a substitute in one concert (March 2025) to play under the current, internationally acclaimed conductor Lori Antounian.

One evening in September 1994, I was surprised by a phone call from a number I didn't recognize. "My name is Ante Padjen, and I'm the director of the I Medici di McGill Orchestra. You've been recommended as a possible flute player for the orchestra."

Yikes, I thought, how did this come about? I hadn't played in an orchestra for over 20 years. Then I remembered that in the previous year, I happened to meet ophthalmologist Michael Flanders who told me that he plays flute in the I Medici Orchestra, an amateur group known as "the doctors" orchestra'. Shortly after our meeting, Michael later came to my home for an evening of duet flute playing, which I really enjoyed.

Dr. Padjen asked me to call the conductor, Wanda Kaluzny (Photo 27.1), to arrange an audition. Wanda was straightforward "No need to audition," she said. "If Michael

L. Mysak, *Adventures in Climate Science, Ocean Waves, and the Flute*, Springer Biographies, https://doi.org/10.1007/978-3-032-19848-8_27

Photo 27.1 Directors and conductors of the I Medici di McGill Orchestra. Upper half, L to R: Tom Samek, Ante Padjen, Lori Antounian (current conductor), and Alexandre Sheasby. Lower half, L to R (past conductors): Wanda Kaluzny, Iwan Edwards, Gilles Auger and Véronique Lussier (Photo courtesy of Alexandre Sheasby)

recommended you, just come to the next rehearsal with your flute and try out with us."

Next week I showed up at the Palmer Theater, the rehearsal room in the McIntyre Medical Building of McGill and took my place beside Michael in the flute section of the orchestra. I was nervous as I had to sightread three or four pieces that evening. But I managed okay, and after the second piece, Wanda gave me a smile, suggesting that I made the grade. "You'll be fine," she said at the mid-rehearsal break. "Welcome to the orchestra."

That was over three decades ago, and until May 2024 I held on to my flute position in the orchestra. My three decades with I Medici di McGill have been a wonderful ride of musical and emotional experiences. I've played under five very different conductors and have gotten to know a wide range of repertoire. And it has been inspiring for me to be able to keep up the flute for so many decades. I've also met colorful and fascinating fellow musicians as well as several outstanding concerto soloists.

Right from the beginning, Michael generously suggested that we alternate in playing the first flute part in some of the pieces. Playing first flute is usually very challenging, and since this part often carries the melody, it is exposed. While the audience may not notice any mistakes, the conductor and fellow musicians sure do! As I got older, I found it more stressful to play the first flute parts, but I have also enjoyed the feeling of accomplishment when I've done a good job.

27.1 Founding and Financing the Orchestra

The I Medici Orchestra was founded in 1989 by Ante Padjen, a then professor of pharmacology at McGill. He proposed to Dr. Richard Cruess, then medical dean, to establish a string chamber group of about 15 players made up of individuals connected to the Faculty of Medicine. Ante argued that an orchestra would offer Montreal audiences the healing power of music [1] through concerts benefiting the healthcare community and local charities. Initially, most players in the orchestra were either medical professors or students. It is amazing how many talented musicians are embedded in the health sciences. Remarkably, the I Medici Orchestra is the only doctors' orchestra in Canada.

Wanda Kaluzny, who already founded and conducted the Montreal Chamber Orchestra, was the first conductor of I Medici, and she continued with us until 2000. Winds and brass were added to the orchestra in the year before I joined, and the string section was expanded. Today there are over 50 players in I Medici, making it a full symphony orchestra. Some of the members even have degrees in music and work in other professions besides medicine. The orchestra often hires a few professional musicians or McGill music students to fill in gaps in the ensemble. There are generally three concerts a year, and after each we have a potluck celebratory party. This party breaks the ice between the conductor and the orchestra members and is a great way to relax and get to know each other.

It costs money to run an orchestra, and in some years, we operated in the red. We pay the conductor, give a modest honorarium to the soloist, rent music and a concert hall, pay for a few professionals, and pay for advertising, recording and the printing of tickets and programs. Although orchestra members are asked to pay "dues" via a charitable donation to the medical faculty, the amount collected this way together with the proceeds from the ticket sales rarely cover all our costs. We quickly learned that some concert venues are more expensive to rent than others (e.g., Oscar Peterson Hall at Concordia, expensive, versus église Notre Dame, a bargain). Now we are very happy to play in the Church of St. Andrew and St. Paul in downtown Montreal.

Despite several fundraising and promotional efforts, the orchestra has always been under financial stress. Thus, starting around 2016, the orchestra adopted a new mode of operation: "You pay, and we play." This means that a health-related charity wishing to raise funds organizes the advertising, ticket sales and donations for a concert. The charity keeps the net proceeds from ticket sales after paying the orchestra's expenses for that concert, which generally add up to around $8000. Some of our past benefit concerts have aided the Montreal Association for the Blind, the Mackay Foundation (for disabled children), Syrian refugees, the Médicins sans frontières in Sudan, Nova Montreal (formerly the Victorian Order of Nurses of Montreal), the Bell Fund for Cancer, and the Cedars Cancer Foundation. We have also given a concert to raise awareness of Parkinson's Disease and two concerts for the medical benefit of the people in Ukraine affected by the brutal invasion by Russia in February 2022.

27.2 Conductors of the Orchestra

Wanda was an easy-going conductor, and playing in the orchestra in those early years was a much-needed time of relaxation for me. Because we are an amateur orchestra, she did not expect perfection. When I joined the orchestra, I was both director of a climate centre at McGill (Chap. 17) and president of the Academy of Science of the Royal Society of Canada (Chap. 19), both of which involved a lot of travel. I really looked forward to the Tuesday evening rehearsals as a time to escape from my demanding scholarly work. One highlight for me during my second year with the orchestra was teaming up with Michael Flanders to play a Vivaldi Concerto for two flutes. This was the first time I was ever a soloist with a symphony.

The renowned choral conductor Iwan Edwards (Photo 27.1) took over as the maestro in 2000 and continued until 2014. He was trained as a violinist in the UK, but after coming to Canada in the 1970s, he spent most of his career leading choral groups in eastern Canada. When in his early sixties, Edwards decided that he would love to return to his roots as a string player and conduct an orchestra. With his no-nonsense style of conducting, he brought us to another level. Sometimes he was tough on those who could not count or play in tune. Since I sometimes came in a bar late, he once said to me, "Lawrence, I don't know what you got your Order of Canada pin for, but it sure wasn't for counting." Embarrassed as I was, I loved the retort by our bassoonist and psychologist professor Jim Katz. "So, Iwan," he said, "I guess you got the Order of Canada because you can count!"

After Iwan retired from the orchestra, Gilles Auger (Photo 27.1), a professor of conducting in the Quebec Music Conservatory, took over in the fall of 2014. He has a master's degree in conducting from the Julliard School of Music in New York and studied under the late Leonard Bernstein. He has conducted many orchestras in Canada, the USA, and Europe, and he has been the long-time conductor of another amateur group, the Lévis Symphony Orchestra near Quebec City. His enthusiasm was infectious, and he worked very hard with the I Medici strings to bring out a more professional sound. At concerts he conducted with no score, which meant that we had to be continuously on guard for subtle changes in tempo and volume as we played. “When you don’t understand the Italian markings in the music,” he would tell us, “They mean ‘look at me’.” Often, I felt very uplifted after our concerts under Gilles’ baton.

When COVID-19 hit us in the spring of 2020, the I Medici Orchestra stopped practicing and performing. In the summer of 2021, however, a small string group from the orchestra gave a concert online, and in the fall a concert of mostly string pieces was presented to a live audience once again. In the spring of 2022, Gilles stepped down as the conductor, and we were then very lucky to hire a young woman conductor to lead us in a benefit concert for Ukraine. Véronique Lussier (Photo 27.1), who has a doctorate in conducting from the Université de Montréal, cheerfully and competently led us through the Grieg Piano Concerto and Beethoven’s 8th Symphony. For me it was great to play under a conductor who is so positive and encouraging. Lussier continued with us until fall 2024.

In December 2024, Lori Antounian (Photo 27.1) became I Medici’s fifth conductor. Like Lussier, Antounian also has a doctorate in conducting from the Université de Montréal. Besides having founded in 2009 her own orchestra in Montreal, the Imperial Orchestra, Maestra Antounian has conducted two major orchestras in Argentina, one in Italy, and the Armenian National Philharmonic Orchestra. With this wealth of experience, I am sure she will be a great inspiration for the I Medici Orchestra.

27.3 Impact of the Orchestra on Me

Playing in the I Medici di McGill Orchestra has certainly kept me feeling “young at heart”. It’s fun talking to the students in the orchestra who study a wide variety of subjects and come from many different locales. Years ago, when as a young teenager I had to decide between evening hockey and band practices in Edmonton, my dad cleverly said, “You’re not likely to be playing hockey at 50, but you might still enjoy playing the flute in an orchestra.” He was so right. Prior to the I Medici Orchestra, I have played flute in the Strathcona Composite High School Orchestra and University of Alberta Symphony in Edmonton, and in the Cambridge Town Orchestra in England, and in the New Westminster Symphony in British Columbia.

Ante often talks about the healing power of music and its role in helping seniors adjust to aging. For the past 10 years, I’ve had a lot of fun giving flute-piano recitals of light classical and popular music in several seniors’ residences in Montreal (Chap. 28). The audiences are captive and very appreciative. For many years, I teamed up with one of my former PhD students, Marc-Olivier Brault who had an ARTC diploma in piano from the Toronto Conservatory. Unfortunately, COVID-19 interrupted this rewarding activity. However, starting in 2022, I have been able to continue these recitals, now with accomplished pianist Karen Boeckh, my wife’s daughter-in-law. She introduces some romantic French songs into our programs, to go along with the classics.

Over the years, I continually asked myself "When will I stop playing in I Medici?" Many years ago, I thought of 80-year-olds as having wrinkled faces and cracked lips who surely could not play wind instruments. I'm lucky that at 85 in 2025, my lips have not yet cracked, and I can still produce a warm tone on my flute. I still have good lungs and agility with my fingers. But in spring 2024, I decided, after 30 years in the orchestra, to retire on a "high note". I informed Ante Padjen and Veronique Lussier of my decision just before the May concert. Much to my surprise and delight, both my children, Paul and Claire, came to this farewell concert (Photo 27.2).

Photo 27.2 Me with my daughter Claire and son Paul, and my wife Janet Boeckh after the May 12, 2024, concert in the Church of St. Andrew and St. Paul in Montreal. While Claire had to only travel two hours by car from Sherbrooke, QC to get to my farewell concert, Paul had a 12-hour journey from Prince George, BC; he arrived at 2 am in the morning of the concert. Fortunately, we played some of his favorite pieces at the concert, including Elgar's Pomp and Circumstance Overture

27.4 Coda

Everyone in the orchestra is amazed that Ante Padjen has been able to keep I Medici di McGill afloat for over 35 years. He has had boundless energy and enthusiasm for the orchestra, and he has many contacts in the health science and musical worlds. Most of us believe that the orchestra would not survive without his leadership; he can get things done, even if at the last minute. He is fortunate to have a dedicated executive to help him manage the orchestra. This turned out to be a very valuable asset when Ante suffered a severe heart attack in December 2024 while visiting a daughter in New York. He did return to Montreal the following month for intensive medical care. Unfortunately, Ante did not make much headway with his healing, and he passed away on November 3, 2025. Thus, an era for the I Medici Orchestra has now ended.

Fortunately, the concert master Alex Sheasby, together with executive members Tom Samek, Michael Flanders among others (Photo 27.1), were able to organize the March 2, 2025, concert conducted by Maestra Lori Antounian. As fate would have it, at the last minute I was called in as a substitute for second flute for this concert because my replacement (David Weigens) was in Europe during January and February. Thus, I can truly say from this last-minute experience that Maestra Antounian, with her passion and enthusiasm, is indeed an excellent choice as the new conductor of the I Medici di McGill Orchestra.

Acknowledgements It is a great pleasure to thank Janet Boeckh for her helpful comments and Josef Schmidt for his refined editorial assistance. The suggestions for improvements to this chapter from Leon Glass, Michael Flanders, Joyce Blond, Lanny Levine, and Lori Antounian are greatly appreciated.

Reference

1. Levitin DJ (2024) I heard there was a secret chord: music as medicine. W. W. Norton & Company, New York, NY

Chapter 28
Dad Was Right: Flute Beats Out Hockey

Abstract At the age of 12 in 1952, I was in a quandary as to whether I would start playing flute in the Edmonton School Boys Band or continue playing ice hockey. The practices for both activities were on the same night! A wise word from my father helped to resolve the issue. "Many years down the road you may not be able to play hockey," he said. "But you might still be playing the flute." And more than 70 years later, this is indeed the case. While I retired from symphonic playing in 2024 (Chap. 27), I continue to play my flute privately for family members and for seniors in residences, where the audiences are very forgiving and most appreciative.

Over a lunch at the McGill Faculty Club during April 2022, Paul Austin said to his friend Josef, "We have real concerts now." During the main part of the COVID-19 pandemic period (March 2020 to March 2022), the senior residents of Manoir Westmount had no live entertainment. Paul, a resident of the Manoir and former McGill professor, was now very excited that this coming Friday evening, there would be a flute and piano concert in the dining and bar area. I was the flutist and Karen Boeckh, the pianist.

When I retired from McGill in 2010, I stopped teaching but continued to supervise a few talented graduate students in climate dynamics as an emeritus professor. Marc-Olivier Brault was one of those students. He also had a diploma in piano performance from Toronto's Royal Conservatory of Music. Since I was a keen flutist, in 2012, I proposed to Marc that we team up as a flute-piano duo and volunteer to give concerts at seniors' residences in Montreal. He thought this was an excellent idea. However, little did I know just where or how often we would be able to do this.

When I was a young teenager, I had to choose between conflicting hockey and band practices. "You may not be playing hockey in your senior years, but you might still be playing the flute," my dad advised. How right he was. With Marc as an accomplished pianist, I could now do some serious flute-playing again by giving concerts to other seniors.

The idea of spending part of my retirement playing flute at seniors' residences goes back to Vancouver over 40 years ago. In the early 1980s, when I was a professor of mathematics and oceanography at UBC, I loved doing concerts at such residences

L. Mysak, *Adventures in Climate Science, Ocean Waves, and the Flute*, Springer Biographies, https://doi.org/10.1007/978-3-032-19848-8_28

with flutist-colleague, Michael Schultz. As well as playing flute duets, the two of us joined up with pianist Mary Mysak, my then wife, to play trios. My sister Helen, who had a lovely soprano voice, also took part in these recitals. I still remember how much joy we brought to the folks in these residences. After a concert, many came up to us and asked when are you coming back? Unfortunately, I had to quit this activity after a couple of years as I was overwhelmed with teaching and leading a growing research group at UBC. In 1986, we moved to Montreal (Chap. 15), and I then had the challenge of establishing a new climate research program at McGill (Chap. 17) and had little time for giving concerts.

During the period 2012–14, after many hours of practice separately and together, Marc and I gave nine concerts in five different Montreal seniors' residences: Fulford Residence (now closed), St. Andrew's Residence, Manoir Westmount, Place Kensington, and Westmount One. These residences were recommended to us by or friends, and at first it took a lot of courage for me to approach the management of each place with the offering of a concert. But once we got started with this venture, we were pleasantly surprised at the number of invitations we received. During the first few concerts, we offered a good dose of the classics, Bach, Handel, Mozart, and Beethoven (Fig. 28.1). But these were interspersed with some lighter pieces by Dvorak and Monti (Csardas is a lively gypsy piece).

After two years, we decided that in the future, we would focus on giving concerts at Place Kensington and Manoir Westmount because at these residences we attracted larger and very enthusiastic audiences. This was most gratifying. But a problem arose. Marc was spending more time practicing for these concerts than doing research for his PhD thesis. Clearly, we had to take a break from our "concert career" so that he could complete his PhD program. He finally finished his PhD in the spring of 2017, and thereafter we did four more concerts, two each at Place Kensington and Manoir Westmount. In early 2019, Marc moved to Ottawa for a job in the industrial sector, but according to his fellow student Chris Simmons he is now back in Montreal working as a barista in the gay village. It is a pity that he has quit both science *and* music, at least for the present.

After the COVID-19 pandemic ended, I was thrilled that pianist Karen Boeckh (the daughter-in-law of my wife, Janet Boeckh) was willing to join me in a new flute-piano duo. At the time of writing this chapter (summer 2023), we have given four concerts so far at Place Kensington and Manoir Westmount, two in 2022 and two in 2023. In contrast to the earlier programs with Marc, Karen, and I have added more popular tunes and some folk music (Fig. 28.2). If the Boston Symphony could do this at the "Boston Pops" concerts, so could we. Karen had never given concerts like these before and was quite nervous at the beginning. However, now she has more confidence and is really enjoying this music-making. Sometimes one of us skips a bar or a whole line of music, and we quickly get out of kilter. When this happens, we stop immediately and start again at the beginning of a section. In addition, on occasion I forget which key I'm in and add a sharp or flat that is not in the musical score. But I just keep going and hope that the audience doesn't notice. The seniors in residents are generally very forgiving.

Manoir Westmount Presents…

A Classical and light classical music for flute and piano, and also for solo flute and solo piano

Thursday, November 1, 2012

8:15 pm

Performers:

Professor Lawrence Mysak, flute, and Mr. Marc-Olivier Brault, piano, from McGill University

Program *(but not necessarily in this order)*

Handel -- Sonata No. 4 for flute and piano, Movements 1 and 2

Mozart -- Andante for flute and piano

Dvorak -- Going Home, from the New World Symphony, arranged for flute and piano

Schubert -- Serenade for flute and piano

Debussy -- Le Petit Berger for flute and piano

Bach -- Sonata for solo flute, Movements 3 and 4

Monti -- Csardas for solo flute

Bach -- Chromatic Fantasie for solo piano

Beethoven -- Pathetique Sonata for solo piano, Movements 1 and 2.

Fig. 28.1 Program for the third concert by the Mysak-Brault duo at Manoir Westmount, on November 1, 2012. Note that we each also performed some solo pieces, to give the other a chance to rest

Spring Afternoon Concert

with

Lawrence Mysak, flute,
Karen Boeckh pianist

Wednesday, April 5, 2023
3:15pm
Auditorium

Lawrence Mysak is an Emeritus Professor in the Department of Atmospheric and Oceanic Sciences at McGill University. He is widely known for his work in applied mathematics and the dynamics of climate change. In addition, Lawrence has a Certificate of Arts diploma in flute performance from the University of Alberta, and for the past 30 years he has played flute in the I Medici di McGill Orchestra

Karen Boeckh is an enthusiastic amateur musician (piano, accordion, and flute). Her first love is classical music, but she also enjoys participating in impromptu folk music concerts at home with family members, who sing and play guitar and mandolin. Moreover, Karen is a part-time English as a Second Language teacher who loves travelling, outdoor activities and learning new languages.

Program

Flute and Piano Duets
Handel.......Largo, Allegro, and Gavotte
Mozart................Andante
Faure ....Sicilienne

Solo Piano
Piaf.............La Vie en Rose
Gershwin...........Summertime

Solo Flute
Bach.........Sarabande and Bouree
Dvorak..................Goin' Home
El Condor Pasa ....Folk Tune
Lloyd Webber.......Memory (from Cats)

Flute and Piano Duets
Greensleeves...........Folk Tune
Steckmest...............Santa Lucia
Monti.............Csardas

Fig. 28.2 Program of the third concert by the Mysak-Boeckh duo on April 5, 2023, at Place Kensington, Westmount

Since retirement in 2010, I have often been asked to play at a friend's birthday party or at the memorial service of a former colleague or family member. For a party, I usually play Monti's Csardas as this is very uplifting and leaves the audience smiling (see Photos 28.1 and 28.2). For a memorial service, I play Massenet's Thaïs or Fauré's Sicilienne, both of which are truly soulful.

During January 2023, Janet and I spent four weeks enjoying the sun, beaches, and food in Bucerias, Mexico, which is just north of Puerto Vallarta. At 5 pm each evening, we had happy hour around the swimming pool patio. Before we left for Mexico, Janet suggested that I take along my flute and some music, in part to keep me in shape for the upcoming I Medici di McGill Orchestra concert, but also to serenade her at sunset. The guests at our lodge, Casa Gardenia, heard me playing the flute and asked whether I might play for them. In the end, I gave three half-hour patio concerts of both classical and popular music, and some guests even invited their friends. The folks especially liked my commentaries on some of the pieces as many had not heard much classical music before. I was, however, at first taken aback by one question after the first concert. "You've been playing the flute for 70 years?" Mark, from Edmonton, asked. "Yes," I said cautiously. "Then why do you still need paper?" At first, I didn't get what he was trying to say. Then I realized that he was

Photo 28.1 Me playing Monti's Csardas at the Faculty Club at the 85th birthday party of engineering colleague John Jonas during 2017. Photo courtesy of Paul Austin

Photo 28.2 John Jonas (left) enjoying my rendition of Csardas at his 85th birthday party together with three of his granddaughters. Photo courtesy of Paul Austin

referring to the music sheets on my stand. By now, he thought that I should be able to play every piece by memory.

"When I was young, I did memorize my pieces for a concert," I told Mark. But today I admit that my memory is not what it used to be! But for the second concert I learned by heart the haunting South American tune El Condor Pasa.

I often wonder how long I'll be able to continue playing the flute. But Dad was right—I can still toot away in my 80s.

28.1 Epilogue

After this chapter was written (summer 2023), Karen Boeckh and I gave three more concerts, the last one in spring 2024. In 2026 I started again giving concerts, now as a solo flutist.

Acknowledgements I thank Janet Boeckh for her comments on the first draft of this memoir, Olivia Marino for inserting the photos, Karen Boeckh, Doreen Friedman and Dan Wayner for their input, and Josef Schmidt for his insightful critique.

Chapter 29
The Country Cottage and Grandchildren

Abstract In January 2004, my then wife Mary and I purchased a three-bedroom year-round cottage in the municipality of Dunham, Québec, about a one-hour drive east of Montréal. The cottage has a wrap-around deck adjacent to an above-ground swimming pool and is surrounded by 12 acres (5 hectares) of pines and hardwoods. Mary I and spent weekends there and had many visits from friends, family (especially the children) and my graduate students. We also befriended a local farmer family. After Mary died in 2011, my children thought I might sell the cottage. Happily, I did not do this, and my wife Janet and I have been able to spend a great deal of time there. We have also hosted the visits of family and friends at the cottage. In addition, we have enjoyed cross-country skiing, cycling, walking in the woods, swimming in the pool, dining in fine restaurants, wine tasting in the vineyards, and summer outdoor concerts. We also do minimalist gardening. We are lucky to have help with maintenance work for the cottage and surrounding forest. As long as we can drive and cook our own meals, we will continue to spend at least half of our time at this oasis.

"What are your three financial goals?" asked Nick Bakish, my then newly acquired financial advisor in fall 2003. After a pause, only two came to mind.

"I'd like to endow a chair in climate research at McGill," I replied. Nick's face drew a blank. I then recalled a note I found 30 years ago written by my then wife Mary in which she expressed her long-term wish: "My own beautiful acreage and house in the country by a lake, or by the ocean," she wrote. So, I next told Nick that my wife always wanted a country cottage.

"You're 63. What are you waiting for?" he retorted. "Think about where you'd like to be in the country and how much you want to spend."

That evening my wife Mary phoned her realtor friend Jackie, who soon found us an agent near Sutton in the Eastern Townships (Estrie, in French), about one hour east of Montreal and just north of Vermont. Mount Sutton is a well-known nearby ski resort, and the land around it has attractive forests, lakes, rivers, and family farms. After our arrival in Quebec in 1986, Mary and I had often spent summer holidays in this area.

L. Mysak, *Adventures in Climate Science, Ocean Waves, and the Flute*, Springer Biographies, https://doi.org/10.1007/978-3-032-19848-8_29

Photo 29.1 Wintertime view of the cottage on McCutcheon Road, Dunham, QC

Next day we were talking excitedly to Peter Reindler from Royal LePage Realty. "We want a quiet, forested piece of land with a low maintenance house in good repair," we said. Within a day Peter sent us six listings. Only one looked of interest to us.

On a cloudy wintry day in January 2004, we quickly fell in love with a 12-acre (5-hectare) property at 1861 Chemin McCutcheon, in Dunham, about a 10 min-drive from Sutton. On it sits a beige 1980s bungalow surrounded by tall white pine, ash, and maple trees (Photo 29.1). There is also a swimming pool next to the back deck.

The kitchen window features a view of the Yamaska River valley and the foothills of Mount Sutton (Photo 29.2). And inside, natural light from five large windows floods the open living-dining-kitchen area. Three bedrooms provide enough space for visiting family members and friends. The partly finished basement contains a wood-burning stove that heats the whole house. We could already imagine the fragrant smell of burning maple and ash.

We immediately accepted the counteroffer from the owner two days after our first visit to the cottage. We didn't want this place to slip away from us, as Peter told us there was another offer on the table. "I'm thrilled we managed to buy this," Mary said.

I still remember our first day at the cottage in early May 2004. The maple leaves were translucent yellow-green, and the birds were chirping constantly. And the aroma of the neighbor's freshly manured farm field brought memories of my youth when I visited the two grandpa farms in Saskatchewan. My dad came to visit us in Montreal in late May 2004, and we took him to the cottage for lunch (Photo 29.3). Because of the smells from next door, he always referred to our cottage as "the farm."

Photo 29.2 View of the Yamaska River valley from the kitchen window of the cottage

Photo 29.3 Dad aged 97 sitting on the cottage deck digesting lunch, in May 2004. He died in April 2007 at age 100, and we buried some of his ashes in the woods, just below the back deck

Photo 29.4 Mary with our daughter Claire on the back deck of the cottage (summer 2004)

Mary and I spent most weekends relaxing at the cottage, often with our adult children (Photo 29.4). In August 2008, Claire married Dennis Taylor on his family farm near Sherbrooke, QC, and in 2010 their first son, Neil Taylor, was born. They often used to spend weekends at the cottage, and we in turn visited them on their farm (Photo 29.5).

The cottage was also a place where I reviewed my graduate students' theses and revised their papers for publication. We had week-long holidays there in summer. We enjoyed eating on the wrap-around deck shaded by pines and maples, and on hot humid days we splashed in the cool water of the above-ground pool next to the deck. For many years, we were lucky to have had Rodney Bear, a Cree from Saskatchewan, help with the opening and closing of the pool in spring and fall. Until 2019, he had also been our handy man and a golf partner for me.

The location of the cottage is very convenient. It's about a one-hour drive from the east end of Montreal's Champlain bridge. Five minutes away is Cowansville, where there are many restaurants, food stores and wine shops. And in less than 15 min, we are strolling down the friendly streets of the town of Sutton, below the ski slopes.

Mary enjoyed spending time by herself at the cottage. She meditated above a cliff at the rear of the property. Early one evening in summer 2004, a cougar slowly ambled by on a trail below. Fortunately, Mary was downwind of this majestic cat, and thus it was oblivious to her presence. I'm still waiting for the cougar's second appearance.

Every summer, Mary and I hosted my graduate students at the cottage. Some preferred to take it easy and simply enjoy a beer on the back deck (Photo 29.6), whereas others liked to play in the pool (Photo 29.7), or have a game of volleyball, or take a walk in the woods.

Photo 29.5 Claire Mysak and Dennis Taylor with their son Neil Tayor (15 months) enjoying apple picking near their farm (October 2011). When Neil was about five, he roared with laughter when I told him, "I went for a swim in our pool with my 'nothing on'." Neil also liked to swim in the pool

Photo 29.6 Relaxing on the back deck of the cottage with some of my graduate students. L to R: Ann-Sophie Cochelin (France), me, Mary, Denise and David Holland (Newfoundland and Labrador), and Andrés Antico (Argentina) (summer 2004)

Photo 29.7 A very competitive game of beach ball in the pool at the cottage. L to R: Bruno Tremblay (Canada), me, Jan Sedláček (Switzerland, with back to the camera), Geoff Lemieux (Canada), and Alexandra Jahn (Germany) (summer 2006)

Sadly, Mary died suddenly in December 2011, after having enjoyed eight wonderful summers there. She especially loved lounging on the massive deck behind the house with a gripping novel in her hands. Some of her ashes are buried under a white pine near the pool.

My children, Paul, and Claire, thought I would sell the property after their mom's death. Instead, in January 2012 my daughter helped me install a 48-inch TV in the living room of the cottage. That month, at 4 a.m. one morning, I got up to watch live, the nearly six-hour epic tennis match between Djokovic and Nadal at the Australian Open. After the match, I wondered—will this be my future, living as a couch potato? My destiny proved to be otherwise.

That January I met Janet Boeckh at a reception following a *Cutting-Edge Lecture in Science* at McGill (Chap. 21). I asked her, "What's your interest in science?" I forgot her reply, but my pick-up line seemed to work. We started dating later that summer, and we married at the cottage in 2015.

During the past dozen years, we've had week-long visits of Janet's grandson Jonah Boeckh when he was 10 and 11. Also, Claire and her two sons (the second son, Austin Taylor, was born in 2013) have continued to visit us at the cottage. The older son Neil has spent week-long holidays at the cottage when he was seven and eight; he enjoyed working with me building benches, tree houses, and board walks for the

Photo 29.8 Grandsons Neil (8) and Austin (5) jostling for position on the paddle board on Lac Brome, QC (July 2018)

wetlands. Neil and Austin also did water sports in nearby Lac Brome (Photo 29.8). Today they are very active and talented in many other sports, including hockey and golf. They also like to go walking with grandpa in the woods that surround the cottage (Photo 29.9).

Since 2013, Janet and I have gotten to know the neighbors much better and have developed new friendships. We are fortunate that Brent McCutcheon, our cow and sheep farmer friend up the hill from us, is happy to have a contract each year to cut our grass in summer and clear the driveway of snow during winter. We also buy fresh eggs produced by the hens kept by his wife Nina and honey coming out of the beehives of Caleb, Brent, and Nina's son. Caleb's children, Caitlan, and Conner like to earn extra money doing gardening and other minor maintenance work on the house. Finally, since 2020 the opening and closing of the pool each year could not be done without the help of another neighbor, Bruno Isabelle, a retired trades teacher at the local high school. He was recommended to us by Rodney Bear, our former handy man.

Now in 2025, I've owned the cottage for over two decades, and because of the help we get from the neighbors, Janet and I enjoy spending much time there each passing year. And I am lucky that my son Paul, who has now settled in Prince George, BC always enjoys splitting wood whenever he visits the cottage (Photo 29.10). In

Photo 29.9 Resting on a bench at the edge of the cottage property. L to R: Neil (13), Claire, me and Austin (10) (October 2023)

November 2019, just prior to the COVID-19 pandemic, Paul was fortunate to meet Allisha Clemetson in Prince George, where they both have partnerships in different engineering businesses. Paul and Allisha married at the cottage in 2022, and their son Julian Mysak was born in December 2022 (Photo 29.11). Paul, Allisha, and Julian visited us at the cottage over Christmas 2023, when Paul did more wood splitting! Paul and Allisha also now have a daughter, Sophie Mysak, who was born in July 2024 (Photo 29.12).

Photo 29.10 My son Paul splitting wood for the basement stove in the cottage

During the first year of the pandemic (2020), Janet and I spent most of our time at the cottage, reading, playing board games, building benches, and clearing up the forest trails. Today, to help keep in shape, we exercise there in all seasons. We cross-country ski in winter and cycle on the hilly backroads in summer (with the aid of a Swytch battery pack installed on our bikes). Janet and I also do aquafit exercises in the pool. We take daily walks, and I still play some par-three golf in Sutton (and Janet is an excellent caddie who keeps my score honest). In the summer evenings, we admire the pulsing fireflies as we walk up the hill to marvel at the constellations overhead.

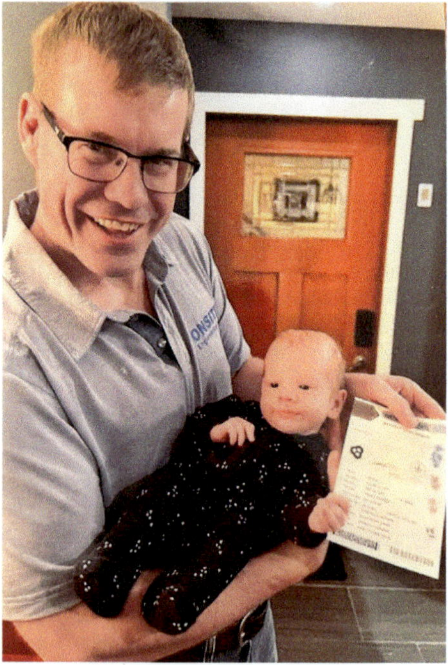

Photo 29.11 Paul with one-month-old son Julian Mysak, holding the official birth certificate (January 2023)

We sometimes wonder how long we'll be able to keep the cottage. Being in quite good health, with valid drivers' licenses, and very happy with each other, we keep our fingers crossed that we'll have many more years here.

Photo 29.12 Halloween time in Prince George, BC for the Paul Mysak family. L to R: Sophie (3 months), Allisha, Paul, and Julian (22 months) (October 2024)

Acknowledgements This chapter is a much-expanded version of a story I wrote for an assignment in the Thomas More Institute (TMI) Memoir Writing Course that I took in winter 2018. I am very grateful to Pauline Beauchamp, and Karen Nesbitt, the course instructors, for their feedback and suggestions for the improvement of this memoir. I also thank Janet Boeckh for her suggestions which helped to improve this updated version and Claire Mysak for inserting the photographs. The comments of Josef Schmidt are gratefully appreciated.

Chapter 30
Cycling Along the Mosel River Valley: Combining Business and Pleasure

Abstract In October 2016, Janet and I took our first cycling trip along a river in Germany, the Mosel. It was a prelude to a workshop that I would be participating in later at the Université du Luxembourg. Thanks to the Mosel-Rad-Touren company, we were able to enjoy a self-guided cycle trip along the river from the historic Roman city of Trier to Koblenz, where the winding Mosel empties into the Rhein. We cycled for five days, averaging about 40 km/day, and each night we stayed in a delightful Gästhaus, often in a local winery which produces fruity Riesling wines. Along the way, we took stops for coffee and pastry and had picnic lunches in the vineyards, which extend down steep slopes to the edge of the Mosel. Having so enjoyed this cycling experience in Germany, we organized two more similar trips, one in 2018 along the Spree, which flows northward to Berlin, and another (for seven days) in 2019 along the Elbe, from Dresden to Magdeburg. Sadly, the COVID-19 pandemic put an end to these lovely holiday adventures.

Our first bike trip in Germany happened because of an invitation to visit Luxembourg. In 2015, Harry Bryden (University of Southampton oceanography professor) and I were asked to write a chapter on ocean circulation for a book on global and climate change, envisioned as a special publication of the International Union of Geodesy and Geophysics (IUGG). The book would contain 30 chapters dealing with various aspects of these changes, and their impacts on weather, urban environments, food security, and geophysical hazards. The 50+ authors were invited to a two-day workshop at the Université du Luxembourg in Belval, Luxembourg in October 2016 to coordinate and integrate our efforts so that the book produced would be a foundational document for the new ISC program "Future Earth.".

Since my travel and living expenses for the workshop would be paid by IUGG and the Université du Luxembourg, my wife Janet asked, "Why don't we top up this meeting with a bike trip?" Online she quickly discovered the Mosel-Rad-Touren company that organized bicycle (= Rad in German) tours along the Mosel River valley, a region in western Germany long known for its fine Riesling wines. We learned that this company could arrange a personalized five-day bike trip for the two of us between Trier and Koblenz (see Fig. 30.1) before the workshop. Since we both

L. Mysak, *Adventures in Climate Science, Ocean Waves, and the Flute*, Springer Biographies, https://doi.org/10.1007/978-3-032-19848-8_30

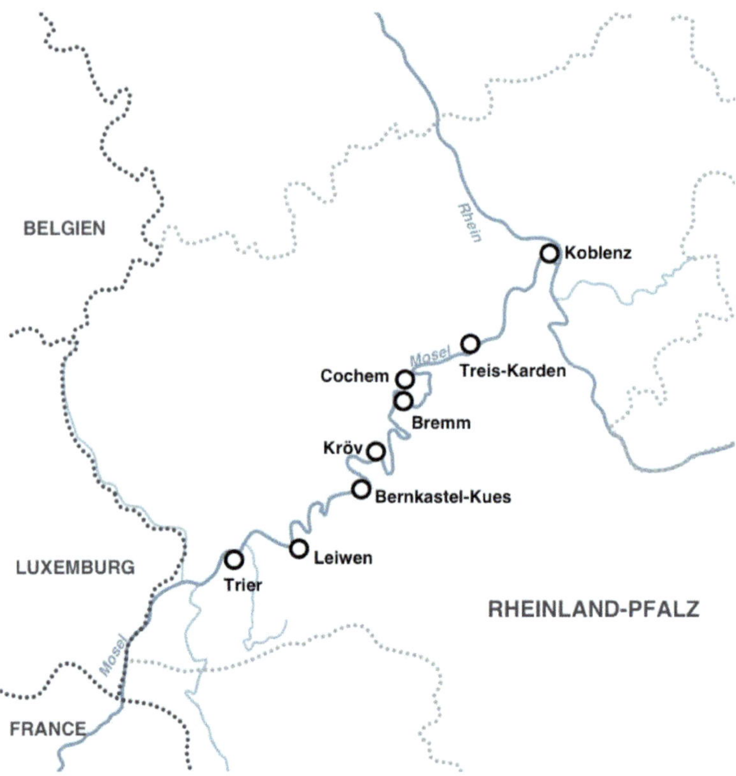

Fig. 30.1 Map of the Mosel River cycle route between the ancient Roman city of Trier (lower left corner) and Koblenz (upper right corner), where the Mosel flows into the Rhein River. After the first night in Trier (Oct. 14, 2016), we spent nights in Leiwen, Kröv, Bremm, Treis-Karden, and Koblenz. (Map drawn by Hannah Blumenthal.)

speak some German, we did not feel the need for a guided tour. However, we were warned that mid-October is late for a cycle trip because of the cooler weather at this time of year. For 900 Euros (1350 CAD) each, we would be able to rent two bicycles from the company, stay in three-star hotels, and have our luggage transferred each morning to the next hotel. We would cycle a total distance of 205 km over the five days, averaging 41 km/day. Since the cycle path was mostly built on an old railway bed along the river, the gradients would be very gentle, which meant that each day's cycle between hotels could be done in about three to four hours. This would give us ample time for coffee and kuchen breaks and side trips to see the many historical sites.

Our adventure began with an overnight AC flight to Frankfurt, where we caught a fast train to Koblenz. There we transferred to the milk-run train to Trier, which had about 20 stops along the Mosel River. We were very tired by this time and did not appreciate an unruly immigrant family of four young boys, whose mother seemed detached and whose father was continuously on his mobile. We had to move to a quieter carriage to preserve our sanity. We were surprised that no one on the train reacted to this out-of-control family.

On arrival in Trier, we discovered that our hotel for the night (the Deutschherrenhof) was only a 15-min walk from the station. Since we just had carry-on luggage, we easily made our way to the hotel and checked in with a very friendly owner. We were delighted to be handed a packet from Mosel-Rad-Touren which contained maps, a personalized day-to-day travel booklet, and vouchers for various treats and museums on our route to Koblenz. After dropping off our bags at the hotel, we headed into the city for a light lunch and then went to the cycle shop at the train station to fetch our bicycles and panniers, which contained repair kits and locks. The bikes were sturdy and had eight gears, which were easy to change. We were happy to see that the bike chain was completely enclosed in a metal casing and thus would not come loose.

We took the bikes for a spin around the town centre and briefly visited the Trier Cathedral (the oldest Bishop's Church in Germany, founded in the fourth century), the Liebfrau church, and a museum of frescoes by Constantine (recommended by my former graduate student Chris Simmons). We also posed in front of Porta Nigra, the impressive Roman-built entrance to Trier (Photo 30.1). Now very tired, we parked our bikes back in the hotel garage and went up to our room for a much-needed rest. Once refreshed, we headed back to town for a complimentary glass of Sekt (a local dry bubbly wine) at the information booth and found our way (in light drizzle) to a Weinstube (wine bar) near the church. We had a great dinner of lentil soup and sausage, while we chatted with the locals at a shared table. They appreciated our attempts to speak German.

We quickly got to sleep that night, but because of jet lag, I woke up at 2 am. Rather than toss and turn, I studied the maps and bike route that we would follow the next day. A hearty buffet breakfast of fruits, cereals, yoghurt, cheeses, cold meats, breads, and even vegetables were most welcome. But this was far too much food for us at breakfast. We thus packed a lunch from food that most Germans would take as a second helping at breakfast! In fact, at every inn we stayed at, the buffet breakfast was most substantial.

30.1 Cycle Day 1, to Leiwen, 38 km (Saturday, October 15)

It was cool and misty at our 9 am start, but we were not discouraged since the forecast was for sun by midday. I was very excited about seeing the winding Mosel River and vineyard-covered hills in the sun (Photo 30.2).

To reach the "Historical Cycling Trail," we crossed the Mosel River at the Kaiser-Wilhelm Bridge just a block from our hotel. The bike path was right beside the river at first, then diverged inland through an industrial part of the town before joining the river again. We soon saw our first vineyards as the mist thinned out.

We were very impressed by how well the cycling path was marked by signs consisting of the letter M together with two bicycle wheels. Although we were given

Photo 30.1 Janet and me with our bikes at the Porta Nigra, Trier, Germany

Photo 30.2 Mosel River valley showing the vineyards extending down to the river's edge. The bicycle path can be seen along the right side of the river in the bottom left part of the picture

a folded map made of plasticized paper, we hardly ever needed it as we were always close to the river. We also enjoyed the option of crossing the river from time to time as there were usually paths on both sides of it.

At 11 am, we stopped in the medieval town of Schweich for cappuccino and delicious German pastry at a bakery near the restored synagogue. But a minor crisis occurred: after the coffee break, we couldn't find the key to Janet's bike lock! We searched in vain in the bakery, the washroom, and the pathway to the locked bikes. I was about to call Mosel-Rad Touren for help at the bakery when Janet shouted, "Look, someone has placed the lost key on the outside window ledge of the bakery." Whew! What a relief.

After cycling for another hour, we found a picnic bench in the vineyards to have our self-prepared lunch (Photo 30.3). Nearby we saw several ancient wineries that go back to Roman times. At the wine village of Detzem, we exchanged our vouchers for a glass of local Riesling and chatted to the local wine producers, who were impressed that we were doing this cycle tour on our own. This wine break was needed since we still had over 10 km to cover before reaching Leiwen and our next hotel.

We arrived in Leiwen at around 4 pm and had to cycle up a long hill to our hotel, the Gästehaus Alexanderhof, which is part of a small winery. We were breathless. At the hotel, we encountered a large group of lively cyclists who teased us by saying, "You've cycled all the way from Montreal?"

Margarethe, our host, led us to the wine cellar, where we stored the bikes and, to our surprise, were given a bottle of their House Riesling. "How generous," I replied when she refused my offer of a few Euros. "We're now in the middle of pressing the new crop for this year's wine," she said, "and we must make room for the new bottles."

Photo 30.3 Me having a healthy lunch in a Mosel vineyard

Leiwen is one of the largest viticultural communities in the Mosel, and the Roman emperor Augustus spent his holidays here. The two restaurants recommended by our guidebook and Margarethe were full that night, but fortunately we were able to get into the Strausswissenschaft dining house, run by one of the local wineries. What an experience! Open only from July to October, it was full of local inhabitants or former locals who relished the home-style meals and enjoyed eating at large tables with other guests. We met a young lady who worked in Luxembourg (at a very good salary) and a man who did radio music shows in Berlin. Another man treated us to local "schnapps" after our pork BBQ dinner. We went to our hotel fully fed and very happy that our first day of cycling went so well.

30.2 Cycle Day 2, to Kröv, 46 km (Sunday, October 16)

Another generous German breakfast was served by Margarethe and her daughter. The lively group of cyclists also joined us at breakfast; they were pros with spandex shorts and tops. After making our lunch, we started out sharply at 9 am as this was our longest cycle ride of the trip. Once again, the day started very foggy.

We first stopped at Neumagen-Dhron, the oldest wine village on the Mosel, where we saw a replica of a Roman wine ship; it was remarkably small. Next came a visit to Café Helga Heim in Piesport, where we used two more vouchers for chocolate truffles, promptly devoured with strong crème coffee. Soon it was lunch time, this one spent on a bench by the river with very steep vineyards behind us (Photo 30.4). Apparently, the vineyards in the Mosel valley are the steepest in the world.

On this second day of cycling, we noticed repeatedly the primeval layers of heat-storing crushed slate that lay over the ground in the steep-sloped vineyards (Photo 30.5). This foundation gives the distinctive mineral taste of the Mosel Riesling wine. We also observed that the vineyards are always on the south-facing slopes, which the Romans discovered to be the best location for growing grapes in this mild micro-climate. But harvesting the grapes on these steep slopes presents many challenges; in some places small funicular carts with buckets were used to bring the ripe grapes down to the river level (Photo 30.5).

After lunch the skies completely cleared, and in Bernkastel-Kues we took a one-hour boat tour up and down the river (Photo 30.6). On this tour, we saw many commercial ships on the Mosel. This explains why we saw very few people swimming in the river. However, on the shores of the river, we did notice several camping and caravan sites, where people were enjoying the mild fall climate.

Bernkastel-Kues (BK) is a popular tourist destination for Germans. There are many fancy shops and restaurants here, as well as attractive four- and five-star hotels. After dodging the crowds walking along the main street of BK, we were thankful to have chosen to stay in the quieter villages that dot the Mosel every few kilometers. (Originally, we were offered the option of staying in these fancier towns.)

After another hour of cycling, we arrived in the beautiful town of Kröv and quickly found our hotel, the Gästehaus Reichshof. This was also a family-run inn with very

Photo 30.4 Mosel River between Leiwen and Kröv, with the vineyards in the distant hills

welcoming hosts. Kristof, the son of the owner, helped us to our room and took us to the storage shed to park our bikes. We then went to find Marcello's Eiseck (ice cream shop), a must-do visit to this village. And yes, we had two more vouchers for free ice cream cones at Marcello's. We had dinner that night in an elegant hotel restaurant, a five-minute walk from our hotel. Delicious trout for Janet (with a glass of Riesling) and Wiener schnitzel for me. I tried a local dry red wine to go with my meat dish, but it was not nearly as good as the Riesling.

30.3 Cycle Day 3, to Bremm, 37 km (Monday, October 17)

Overcast and cool again in the morning, with rain a possibility. Fortunately, it's a short ride today. We visited a private wine museum, where we learned more about the ancient art of wine making. Also saw an unusual collection of dusty old musical instruments there. I wondered who played them last. We had afternoon coffee at a restaurant which served warm apple strudel, which suited us fine because of the outside chill.

Once on our way we encountered our first rainfall, which followed us all the way to Bremm. After 30 min, we were soaked, but luckily at the Hotel Hutter, we discovered heated towel pipes in the bathroom and were able to dry our clothes. We

Photo 30.5 Slate rock foundation of the Mosel River Vineyards. Note the miniature funicular cart that is used to bring the harvested grapes down from the hill

walked partway up Calmont, the steepest vineyard in the Mosel—a 60° slope—and had a fantastic view of the river valley, now cleared of the earlier rainstorm. We then stopped at a wine bar for local Sauvignon Blanc and met German tourists who always bought a few cases of their local Riesling. The waiter did not recommend our hotel for dinner. Instead, he suggested we try the Hotel Berg for their Hirsch goulash (deer stew) and spätzle. Great idea! A short walk around the little village is all we could manage before hitting the sack.

Photo 30.6 Janet on a boat ride on the Mosel, near Bernkastel-Kues

30.4 Cycle Day 4, to Treis-Karden, 45 km (Tuesday, October 18)

Good weather today right from the beginning; breakfast was excellent again, but for the first time Janet had to pay extra for hot milk for her coffee! Halfway through our cycle we stopped in picturesque Cochem, another very touristy town. Main feature was the Gothic castle perched high on a hill. We could only reach it by climbing up a steep set of stairs. But the view was worth it. Later, while licking our ice cream cones, we took a lovely walk along the river on a boardwalk designed for visitors. Seeing the hustle and bustle of this town confirmed that our decision to stay overnight in smaller villages was a good one.

For the first time, we had trouble finding our hotel for the night—we weren't sure which side of the river it was on: the Treis side or the Karden side. After asking the locals, we finally discovered the Weingut-Gästhaus Otto Knaup on the Treis side. Knaup's charming daughter, about 8, showed us to our spacious room, which overlooked a courtyard and had a view of the old city. This was the most elegant room of our trip. Later, the daughter even sold us a bottle of house wine; she sure knew a lot about wine.

The Knaup family has been producing wines for generations, and next morning we bought their special reserve halb-trocken Riesling to take to the Anderssons, whom we would be visiting later in Zürich. We still had Mosel-Rad-Touren vouchers for three wine tastings that evening at the hotel restaurant; but even with these libations we had trouble finishing the enormous pork schnitzels and large plates of Rösti potatoes.

30.5 Cycle Day 5, to Koblenz, 39 km (Wednesday, October 19)

Our special treat for breakfast was smoked salmon, served by the mother of Matheus, the vintner of the Knaup family. Matheus gave us a very informative tour of the wine-making equipment and wanted to take us on a bus tour to see the vineyards. But since rain was in the forecast, we declined his kind offer and pushed on to Koblenz, the end point of our cycle trip.

Halfway to Koblenz, we discovered a delightful bakery in Kobern-Gansdorf, where we had tasty walnut cake and coffee. We were surprised to be the only clientele in the shop.

Every day we were amazed to find excellent cafés and pastry shops with a large selection of delicious cakes—rich but not too sweet. Germans usually have these in the afternoon, but we liked to indulge in the morning after a couple of hours of cycling.

We arrived in Koblenz before lunch but had trouble getting to the hotel. We could spot its location on the map, but we couldn't see how to cross the many railway tracks which blocked our route. The well-marked bicycle path petered out at this point, and only by trial and error did we finally make it to the Hotel Hohenstaufen, in a modern high-rise building. Since our room was not yet available, we left our bikes at the hotel garage and headed for the scenic Deutsche Eck, where the Mosel flows into the Rhein. After walking about a kilometer, we came to the Seilbahn station where we could take the cable car up to the Ehrenbreitstein fortress on the east side of the Rhein River. We did this after eating our sandwiches in the park near the station (Photo 30.7).

The view from the fortress was spectacular, and the surrounding grounds were beautifully kept. After five days of cycling, it was great to have a vigorous walk in the open air, away from city traffic and crowds of people.

Upon returning to the Deutsche Eck by cable car, we explored the old city of Koblenz, which appeared to be thriving economically. We then returned to the hotel for a much-needed rest. That evening we had our final treat provided by the cycle company —dinner at the Weindorf Koblenz, which is part of a conference center, a 10-min walk from our hotel. Excellent meal of salmon and noodles, but the other guests in the restaurant were very noisy. Across the room sat a large party of Japanese businessmen who were more interested in drinking than eating. We should have had the courage to ask for a quiet table in another room!

Photo 30.7 Janet overlooking the Rhein River from the Ehrenbreitstein fortress

30.6 Day 6, Return to Trier and Trip to Luxembourg (Thursday, October 20)

Early breakfast today as we had to be ready at 8.45 am to meet Ronnie, who would drive us and the bikes back to Trier in the company van. When he arrived, it started raining, and on the drive back it poured! We were so glad to have finished our bike trip yesterday. It took us only one hour to drive to Trier on the highway. On the winding river bike path, we did 20 h of cycling over five days, a much more relaxing trip.

After returning the bikes to the Trier station, we explored the city once more before catching the train to Luxembourg. We arrived at Belval-Université station in mid-afternoon and made our way to the modern Iris Hotel, where all the authors for the workshop were housed. We had an informal meal of sushi near the hotel and got to bed early so that I would be fresh for the workshop starting tomorrow morning.

The two-day workshop was very successful. Authors summarized the contents of their chapters (Harry and I each spoke about our ocean circulation chapter, with me focusing on the paleoceanography part). Then there was much discussion on how each chapter complemented or overlapped with the others. After external reviews of all the chapters, it took two more years before the book was published. I was then the very proud co-author of a chapter with Harry [1]. This turned out to be my last scientific publication.

30.7 Epilogue

Janet and I were delighted to discover that we could travel very well together on a self-guided bicycle trip. We had to search for the right route sometimes and make split-minute decisions as to which bridge to cross and which path to follow. Also, we generally agreed when it was time for water and coffee breaks, and lunch. Janet is the one who always wants to pose questions to the locals when we get lost. I am more of an explorer and like to find the way by myself. I love to follow a map and know exactly where I am. We tease each other about our differences in travel philosophy.

With this successful trip behind us, we decided to take two more cycle trips in Germany. In May 2018, we cycled along the Spree River, from Cottbus to Berlin, over a five-day period. In May 2019, we cycled along the Elbe River, from Dresden to Magdeburg, this time over a seven-day period. We thoroughly enjoyed both trips. Sadly, the COVID pandemic in 2020–22 put a stop to further such adventures.

We also enjoy the prospect of meeting friends after a cycling trip and sharing our experiences with them. After the Mosel trip, we spent two happy days with the Swedish couple Ann and Gören Andersson, who at the time lived in Birmendorf, just outside of Zürich (Photo 30.8). When I was on sabbatical in Zürich during 2000–01, my then wife Mary and I had met Ann and Gören, who were also newcomers to Switzerland. Gören had just taken up a professorship in power engineering at ETHZ.

After the 2018 and 2019 cycle trips, we met with John and Rosemary Johnson, who are retired and live in Norwich, England. I have known this couple since 1971, when I spent a sabbatical at the nearby University of Cambridge (Chap. 11). Since John is an applied mathematician and theoretical physical oceanographer like me, we have many scientific interests and colleagues in common. And, like us, Rosemary, and John also love the outdoors and enjoy a good walk. Sadly, Rosemay died May 8, 2025.

Now a decade after our first cycling tour in 2016, I am happy to report that Janet and I still do a lot of cycling. We prefer to get around the city on our bikes, and we enjoy exploring the Eastern Townships on the country roads around the cottage (Chap. 29). We also participate in Montreal region bike tours organized by the McGill University Retirees Association.

Photo 30.8 Janet and me sharing a meal with Ann and Gören Andersson (on the right) in a Birmendorf restaurant, in Switzerland

Acknowledgements Thanks to Janet Boeckh for keeping a detailed diary of the Mosel cycle trip. With this in hand, I was able to describe many of the specific sites we visited and experiences we shared on this trip. I am also grateful to Janet and Josef Schmidt for their comments on various versions of this memoir. Finally, I thank Olivia Marino for assistance in inserting the photographs.

Reference

1. Bryden HL, Mysak LA (2018) Ocean circulation: knowns and unknowns. In: Beer T, Li J, Alverson K (eds) Global change and future earth: the geoscience perspective, Chap. 12. Cambridge University Press, pp 159–175

Chapter 31
Rewarding Good Work

Abstract Scientists spend many hours doing research and writing up their results for publication. Over the years, prizes, awards, and honors have been established to recognize scientists' good work that is displayed in these publications. To win one of these prizes, etc., someone must make a considerable effort to nominate a colleague for the honor. In this chapter, I describe the nomination process that I followed to have one of my former graduate students win a prestigious Canadian prize for an outstanding climate science book that the student recently published.

"Daddy, you've been working on this letter for a long time. What's it about?" my daughter Claire asks.

"I'm writing a letter to nominate a colleague for a prestigious Canadian prize," I reply.

"Do you get paid to do this?" Claire responds.

"No, why should I?" I answer. "In the past, I have been lucky enough to receive several awards and prizes. And to get each of these honors, someone had to nominate me. So now it's payback time—my turn to nominate others."

Each fall the Canadian Meteorological and Oceanographic Society (CMOS) encourages its members to nominate worthy students and colleagues for one of the Society's prizes or awards—there are about 10 of them. Claire saw me working on a cover letter for the nomination of Victoria Slonosky (Vicky), a former graduate student whom I co-supervised, for the President's Prize of CMOS. This is an annual prize given out at the spring meeting of CMOS for a paper or book of special merit in the fields of meteorology or oceanography. In 2018, Vicky published an excellent book about the weather and climate of the St. Lawrence River region for the past 300 years [1]. It's an amazing piece of scholarship, involving both scientific data and history. In my covering letter, I was explaining why Vicky deserved the President's Prize for this book.

During my 43-year career as a professor, first at UBC (1967–86) and then at McGill (1986–2010), I spent many hours writing research grant proposals. The money from these grants was used to cover the research costs and stipends of my graduate students who were engaged in various theses projects. Supervising graduate students was

L. Mysak, *Adventures in Climate Science, Ocean Waves, and the Flute*, Springer Biographies, https://doi.org/10.1007/978-3-032-19848-8_31

generally a lot of fun, and I really enjoyed seeing the joy in their faces when they told me of their exciting new discoveries. At other times, a student seemed less enthusiastic about his or her project, and it required a lot of effort to get that student over the finish line, i.e., complete the thesis in time for graduation. When a student showed great talent for research, I would push him or her very hard to write an outstanding thesis. In such a case, it was a pleasure to nominate the student for an Outstanding Graduate Student Prize given out each year by CMOS, and to write strong letters of recommendation for future postgraduate studies or employment. Naturally I was very happy if a student won an award or was offered a great job.

After about a decade at UBC, my published papers and the work of my research group gained national and international recognition. In 1981 Paul LeBlond and I jointly received the CMOS President's Prize. This was awarded for our 600-page treatise *Waves in the Ocean,* which was published in 1978 [2] (Chap. 13). I still remember the thrill I experienced in spring 1981 when we got news of this award. An even greater joy for me was the presentation of the prize at a CMOS conference in Saskatoon, where I was born in 1940. This prize certainly made me feel that all the work Paul and I had done writing the book over a three-year period was worthwhile and highly appreciated by the oceanographic community. That summer I vowed to help other colleagues receive similar recognition for their research work.

In the years following 1981, I have nominated several former graduate students and other colleagues for the CMOS President's Prize. In each case the nomination was for a research publication. However, I had never nominated anyone for a published book. A surprise visit to my office by Vicky Slonosky in the spring of 2019 changed all that. Vicky had graduated with an MSc from McGill in 1996. After doctoral studies in Europe, in which she focused on analyzing historical weather and climate records, Vicky kept in touch about her ongoing research. Vicky came to my office to tell me that she had recently written a book entitled *Climate in the Age of Empire: Weather Observers in Colonial Canada* [1].

"What an achievement," I said. "Tell me more about it."

Vicky explained how she pored over many historical and early weather records (in both French and English) that she found in various archives in towns and villages along the St. Lawrence River. Today we call this "data rescue work." Among her discoveries, one particularly stands out: From the data she showed that in Montreal, the minimum winter temperature has gradually increased over the twentieth century; however, the average summer maximum temperature over the same period has not increased.

The next day I found a copy of Vicky's book in my mailbox at McGill. Later that summer, I read Vicky's book from cover to cover and quickly realized that with the publication of this book, she would be an excellent candidate for the 2019 CMOS President's Prize. Vicky had become a leading expert in the field of historical climatology, and her work has a strong interdisciplinary appeal since she describes in her book many social science and humanitarian aspects of weather and climate.

Since 2005, Vicky has been on disability leave due to complications from an autoimmune condition; in her last permanent position she served as a Research Associate with OURANOS (Consortium for Regional Climate Change and Adaption) in

Montreal. However, she has continued to publish research papers in her field, to give seminars on data rescue and historical climatology, and to serve as a consultant to many environmental and educational institutions. It is remarkable that Vicky found the energy and time to complete her book, which has received the highest praise from international experts in the field. To complete the nomination of Vicky for the President's Prize, I had to obtain several letters of support. I had no trouble getting excellent letters from her colleagues in Europe, North America, and Australia. As I had hoped, in spring 2020 Vicky received a letter from CMOS telling her that she had won the 2019 President's Prize.

Now that I have been officially retired since 2010, I have the time to continue with this type of recognition work. It keeps me in touch with recent research and the scientific accomplishments of my former graduate students and colleagues, who are scattered around the globe. I'm always happy to write letters of support for colleagues in Europe, North America and elsewhere who have been nominated by others for various awards and prizes. In addition, I have made many other nominations of colleagues and former students for honors and awards distinct from CMOS. This includes several successful nominations of scientists for FRSC, Fellowship in the Royal Society of Canada, our national academy of the arts and sciences.

Recently, I was delighted to learn that my daughter made her first nomination of a colleague for an award, namely, "The Women in Insurance Award" given by Insurance Business Canada. "And guess what?" my daughter later said. "She was selected as a finalist."

References

1. Slonosky VC (2018) Weather in the age of empire: weather observations in colonial Canada. American Meteorological Society, Boston, MA
2. LeBlond PH, Mysak LA (1978) Waves in the ocean. Elsevier Scientific Publishing Co, Amsterdam

Chapter 32
Ukraine: Reminiscences and Survival

Abstract Part 1 of this chapter is a revised version of an article I wrote for the McGill University Retirees Association (MURA) newsletter which appeared in summer 2022. The executive of MURA knew that I am of Ukrainian ancestry and wanted a personal account of how I felt about the brutal invasion of Ukraine by Russia on 24 February 2022. When this article was written In April 2022, I hoped that the war would soon be over, and the rebuilding of Ukraine could soon begin. How wrong I was. Part 2 was written in November 2025, and there is still no end to the war in sight. It has been difficult for me to put words on paper for this second part, but I hope the reader will sense my enormous concern, admiration, and support for the Ukrainian people.

32.1 Part 1: April 2022

In April 2022, there were 1.4 million people in Canada of Ukrainian ancestry, the largest such diaspora outside of Ukraine and Russia. I am one of these people, a second-generation Ukrainian Canadian who grew up on the windswept prairies.

My mother was born in 1907 in Gimli, Manitoba to parents who came to Canada from western Ukraine in 1899. (At that time this region, Galicia, was part of the Austro-Hungarian Empire). In 1910, the family moved to Saskatchewan and started farming on better land near Saskatoon. My father was born in 1906 in western Ukraine in the village of Kluventsi, near Ternopil (then also part of Galicia). He emigrated to Canada in 1912 with his mother to join father, who had arrived a few years earlier to establish a farm near Regina. My parents met in Saskatoon during the 1920s when they were at Normal School, training to become country teachers. They married in 1930 and had two children, Helen born in 1932 and me, born in 1940. Both Helen and I spoke Ukrainian before we learned English. While growing up, we were surrounded by Ukrainian family and friends and participated in many activities at the Ukrainian Orthodox church and community centre.

After World War II, I spent many summers on the family farms and learned about the value of hard work and survival on the land. Both sets of grandparents put a

L. Mysak, *Adventures in Climate Science, Ocean Waves, and the Flute*, Springer Biographies, https://doi.org/10.1007/978-3-032-19848-8_32

high priority on education. They established one-room schools in their respective communities and encouraged their children to go on for higher education. Several of their children served in the war. Fortunately, they came home safely.

My late wife, Mary Mysak, and I traveled to Ukraine three times, first as camping-auto tourists in 1972 during the height of the cold war, and then twice after Ukraine gained its independence in August 1991. In 1993, I was invited to give a public lecture on climate change at the fall opening of the University of Kyiv Mohyla Academy (founded in 1615). During this visit, I also gave a lecture on climate physics at the National Academy of Sciences of Ukraine (NASU) in Kyiv. In 2002, my sister Helen joined Mary and me when I was an official guest of NASU. As part of a lecture exchange agreement between NASU and our national academy, the Royal Society of Canada, I gave climate and oceanographic seminars in Kyiv, Sebastopol and Lviv. By this time, my Ukrainian was rusty, but my father, then 95, helped me prepare my introductory remarks (in Ukrainian) for each of my seminars.

My family has been devastated by the Russian invasion of Ukraine on February 24, 2022, and has been following the events there closely. As a Ukrainian Canadian, I can understand why the citizens of Ukraine are defending their country with tremendous tenacity, courage, and resilience. Over the past several centuries, Ukraine has been ruled by Mongolia, Poland, Lithuania, Imperial Russia, and Germany [1]. Only once has it been an independent country and that was for a few years just after World War I [1]. Since then and up to the dissolution of the Soviet Union in 1991, Ukraine was very much under the political and economic domination of Moscow, except for a brief period during World War II when it was taken over by Germany.

When traveling in Ukraine by car in 1972, Mary and I were under continuous observation by the police. Our travel routes were very constrained and talking to any person on the street was discouraged. Much later we learned that locals could lose their job for doing this. I had one frightening experience at a campsite in Chernivtsi (in southwest Ukraine, next to the Romanian border). A local guide showed me pictures of six young adults from Saskatchewan who were camping in the same region and making a nuisance of themselves by talking about politics with the locals. He asked if I knew any of these people. Of course I pleaded ignorance, even though one of the women in the group I recognized as a former girlfriend.

On our second visit in 1993, when Ukraine had now been independent for two years, it was quite evident to us that, like elsewhere behind the old iron curtain, the people had difficulty adjusting to democracy and self-rule. They were challenged with a new financial system and the setting up of nongovernment-run shops and companies. Right after independence, it was hard for people to find work and make a living [2]. It was not easy for us to find good restaurants, so we always carried a box of crackers and nuts for snacks when exploring Kyiv. We also observed that the people were cautious about showing any initiative. However, they were happy to be free of communism and looked forward to a future with stronger ties to Europe. An experience that Mary and I will never forget was at an open-air luncheon under fragrant fruit trees at the country house (the "dacha") owned by the president of the University of Kyiv Mohyla Academy. After a wonderful spread of food and wine, the 25 guests began to sing. Ukrainian folk songs tend to be sad, but there was joy in the way the songs were sung that day. "The indomitability of the human spirit" was how Mary described that moving experience in 1993.

During our third visit in 2002 conditions were notably better. Restaurants and markets were flourishing and much of the drabness of the Soviet era had started to disappear. Conversations with the people on the streets were relaxed. The country was still learning to deal with democracy and corruption, but we felt that Ukraine, with a population of nearly 50 million, was on its way to closer ties with the West. A special highlight for Helen and me was meeting distant relatives in Ternopil (east of Lviv, but still in western Ukraine) at the end of our visit. We met my father's cousin, Rozalia (then about 80) and four generations of her family, including a friendly six-year-old Andrij. Today I believe he is fighting in the war in Ukraine. Before the war started, I was hoping to visit Ukraine again to introduce my present wife Janet to these relatives and visit the beautiful cities of Lviv and Kyiv.

An excellent article in the 26 March 2022 issue of Canada's widely-read newspaper, The Globe and Mail, by Senior International Correspondent Mark MacKinnon gives a detailed account of Ukrainian history during the past 20 years. The struggle with democracy and corrupt politicians continued for some time after independence. An important event in 2004 was the Orange Revolution when the people rallied together in Kyiv to oust President Viktor Yushchenko, who was under the thumb of Vladimir Putin, the president of Russia. It is only with the last two Ukrainian presidents, chocolate tycoon Petro Poroshenko and comedian-actor Volodymyr Zelensky (elected in 2019) that Ukraine has been able to break free from political domination by Moscow and develop closer ties with Europe. This, of course, is one of the reasons why Putin's army invaded Ukraine; Putin wants Ukraine to remain under the umbrella of Russia and be part of a revived Russian empire. The Ukrainian people's staunch defense of their country during the current war has been unexpectedly strong and admired worldwide. This defence has been made possible with the help of extensive military training and weapons received from the West.

If you would like to know how to help Ukraine and its refugees during this difficult time of war, here are some ideas:

On April 3, 2022, the I Medici di McGill Orchestra, of which I was a member until 2024 (Chap. 27), gave a sold-out benefit concert for people affected by the war in Ukraine. The goal of raising ten thousand dollars for Doctors Without Borders/ Médicins sans frontières (MSF) for their humanitarian work in Ukraine was achieved. Donations to MSF in support of Ukraine are welcome any time.

In the McGill Reporter and the McGill Communications newsletter, activities organized by people at McGill which are supporting Ukraine have been listed (e.g., see the newsletter of 29 March 2022). Requests for donations and volunteers are listed in these publications.

Outside of McGill, donations to the Canada Ukraine Foundation, a non-profit charity, are welcome. This is where I made a major donation shortly after 24 February 2022 and continue to support annually.

The Ukrainian Canadian Congress, an NGO, is focussed on developing lists of Canadians who are willing to assist and house Ukrainians, especially families with children, who have been arriving continuously in Canada since the war started. On a personal note, I am delighted to report that one close relative has recently taken a Ukrainian family of three into their home in British Columbia.

I have also learned that there are many other NGOs in Canada which have been helping Ukraine right after independence and have intensified their efforts since the invasion. One example is the Medical Mercy Canada Society (MMCS) which has provided medical supplies and services to villages outside of Kyiv. MMCS is now providing supplies to the front line in the war. My cousin Marlene and her husband Brian have been working with MMCS since the 1990s.

It is hard to imagine how and when the war will end. However, I hope that one day there will be peace again and the process of rebuilding the country will begin, which will take enormous resources—billions of dollars—from Europe and North America. When this happens, it is likely that many of the refugees will be able to return home and help with this work. I even hope that one day the Royal Society of Canada can renew academic exchanges with the National Academy of Sciences of Ukraine.

32.2 Part 2: November 2025

It has been heart-wrenching to watch the war going on for nearly four years. Unfortunately, there is little sign of it ending soon on terms favorable to Ukraine. There has been much devastation of civilian property, schools, medical facilities, and energy and transport infrastructure. Also, Ukrainian cultural institutions such as museums have been destroyed, and open-minded artists and writers have been targets of assassination. Tens of thousands of Ukrainians (both civilian and soldiers) have been killed, and tens of thousands of children have been kidnapped by Russian soldiers. Over 10 million Ukrainians have been displaced, 3.7 million inside the country and 6.7 million abroad (Office of the UN High Commissioner for Refugees). It is estimated that three hundred thousand have sought refuge in Canada; when added to the Ukrainian diaspora, I calculated that over 4% of the Canadian population is now made up of ethnic Ukrainians.

During the past year (2025), there has been a rapid shift to drone warfare, which has enabled Ukraine to destroy many parts of the oil, transport, and military infrastructure in Russia. Ukraine's expertise in drones has also helped to keep the Russian army from making significant advances in the occupied regions of Donetsk, Luhansk, Zaporizhzhia, and Kherson during the past two years. Including Crimea, which Russia illegally annexed in 2014 after the February 2014 Maidan Revolution in Kyiv, about 20% of Ukrainian territory is now under Russian control. (In fact, the Canadian Council of Foreign Relations and others consider the war really started after the ousting of Russian-friendly President Viktor Yanukovych in February 2014 who suddenly reversed his position on having stronger ties with the European Union. This was immediately followed by the annexation of Crimea under Russian soldiers' guise). When I wrote part 2 of this chapter in November 2025, there were no signs of a peace agreement with Russia on terms that are acceptable to both countries. Putin wants Ukraine to cede at the minimum those parts of the country now under occupation by Russian troops. President Zelensky and the Ukrainian people, on the other hand, demand that Russia withdraw unconditionally from the occupied territories and respect the sovereignty of the 1991 borders of Ukraine.

In the eyes of Putin, the root cause of the war (which he calls a "special military operation") is Ukraine's desire to be part of the European Union and a member of NATO. Putin and many Russians do not see Ukrainians as a separate people [3]. Russia has long regarded Ukraine as a vassal [4]. Zelensky's goal is to break the cycle of Russian imperial oppression that has gone on for many generations [5]. Many do not remember that in 1994, Ukraine agreed to surrender its arsenal of nuclear weapons (at the time, the third largest in the world) in return for the US, the UK and the Russian Federation respecting independence, sovereignty and existing borders of Ukraine [6]. Clearly, this treaty has been broken by Russia.

For some time, Ukraine has had strong military and intelligence support for the war from the USA, Europe, and Canada. However, with President Trump now in the White House since January 2025, American support has waned. Consequently, Zelensky has turned to Europe and Canada to create a "coalition of the willing" for sustained future support (Mark MacKinnon, Globe and Mail, 6 March 2025, page A4). In this regard, I find it heart-warming that Canada's ambassador to Ukraine, Natalka Cmoc, has recently agreed to have her appointment renewed because she believes the presence of many western diplomats in Kyiv serves as a deterrent against even worse Russian attacks on the capital. President Andrzej Duda of Poland has recently said [7, Chap. 61]: "My biggest dream is to make sure that Russia does not win over Ukraine. This is absolutely the most important thing for the security of Poland, and for the security of the whole world."

We have had relative peace in Europe since 1945 [3]. This has clearly been broken by Putin's invasion of Ukraine 77 years later. Unfortunately, we do not have a crystal ball to tell us when or how the war in Ukraine will end. I do believe that Ukraine will not stop defending itself. If Russia does permanently gain territory from Ukraine, i.e., win the war they started, there will be no peace and security in Europe and the western world.

32.3 Epilogue

On a brighter note, I was delighted to recently learn from my cousin Marlene that Rosalia's great grandson Andrij, whom I met in 2002, has returned home safely from his mandatory service in the Ukrainian army. He is now happily married and works as an IT expert in Prague.

Acknowledgements Much of my knowledge of the past and current history of Ukraine (up to 2008) has been gleaned from the internationally acclaimed books by Orest Subtelny [1, 8]. The remaining books listed in the references deal with more current issues in Ukraine and are highly recommended reading.

I thank Janet Boeckh for her comments on several drafts of this chapter. I am also grateful to Niky Kamran, Ginette Lamontagne, Marlene Mysak, Mary Feher and Ireneus Zuk for their critique and input.

References

1. Subtelny O (1988) Ukraine: a history. Univ of Toronto Press, Toronto
2. Reva M (2021) Good citizens need not fear. National Geographic Books, New York, NY
3. Ash TG (2023) Homelands: a personal history of Europe. Yale University Press, New Haven, CT
4. Popova M, Shevel O (2024) Russia and Ukraine: entangled histories, diverging states. Polity Press, Cambridge, UK and Hoboken, NJ
5. Shuster S (2024) The showman: inside the invasion that shook the world and made a leader of Volodymyr Zelensky. Harper Collins Publishers, New York, NY
6. Ukraine, Russian Federation, United Kingdom of Great Britain and Northern Ireland, United States of America (1994) Memorandum on Security Assurances in Connection with Ukraine's Accession to the Treaty on the Non-Proliferation of Nuclear Weapons. 5 December 1990, number 52241. See also "Nonproliferation and the Budapest Memorandum" in Physics Today 78 (page 10, September 2025)
7. Woodward B (2024) War. Simon & Schuster, New York, NY
8. Subltelny O (2009) Ukraine: A History, 4th edn. Univ of Toronto Press, Toronto

Chapter 33
Giving Back

Abstract I benefited greatly from scholarships that enabled my university studies and exposed me to enriching experiences abroad during 1957–67. Since then, I have carried a lifelong desire to give back. This chapter describes how, many decades later, I established endowed scholarships, awards, and prizes at McGill (in science, music and religious studies) and at the University of Alberta (in science). In addition to honoring my parents and late wife Mary Mysak, I have created two undergraduate awards in my name to recognize outstanding students studying climate science and music performance, topics very dear to my heart. Altogether, establishing these endowed awards and prizes has given me much fulfillment in supporting future generations of students.

After World War II, Canada and many other countries invested extensively in university education, which included creating bursaries, scholarships, and prizes. I was fortunate to have received several of these scholarships as an undergraduate and graduate student. As I came from a family of modest means—my parents didn't own a home until I was in grad school—these awards allowed me to study outside of my hometown of Edmonton, resulting in experiences that changed my life. The deep impact of this support planted a seed: one day, I resolved, I would offer similar opportunities to others.

As an undergraduate at the University of Alberta (U of A), I received scholarships from the Canadian Legion, the Alberta Government and General Motors of Canada. The Edmonton Ukrainian Community also gave me a special stipend to study flute performance at the famed Aspen School of Music during summer 1958 (Chap. 26).

For my first year of graduate studies, I was awarded a Rotary Fellowship for International Understanding to do an MSc in mathematics at the University of Adelaide during 1962–63 (Chap. 5). One of the conditions of this award was that I would travel throughout Australia giving talks to Rotary Clubs about my home country and about my experiences living in Australia. To prepare for this activity and give me confidence, I took private public speaking lessons with a trainer for radio announcers. My time in Australia was a transitional year and an enriching experience. I shifted

L. Mysak, *Adventures in Climate Science, Ocean Waves, and the Flute*, Springer Biographies, https://doi.org/10.1007/978-3-032-19848-8_33

my research interests from general relativity to fluid mechanics, and I met graduate students from many countries.

To support my subsequent PhD studies in applied mathematics and geophysical fluid dynamics at Harvard, I received a graduate teaching assistantship, scholarships from both Harvard and the National Research Council of Canada, and a research assistantship from the US Office of Naval Research (ONR) (Chaps. 6 and 7). Having benefited from all these awards and stipends, both private and government, I hoped that one day in return I would be able to provide support for future generations of students. I certainly appreciated the generous support that I received for my university and music studies.

Throughout my four-decade career as a professor at UBC and then at McGill, I often gave modest donations to the U of A, McGill and Harvard for general scholarship support. With two young children (Paul and Claire) during my earlier years as an academic, I was not in a position financially to establish any major awards. To increase my income and build up my net worth, I did contract work for the ONR and invested in real estate (rental properties) in the late 1970s and 1980s. By the early 2000s, I was able to consider endowing a major scholarship. To honor my parents who instilled in me a love of learning and had given me so much support and love, in November 2006, with the proceeds from the sale of a rental house I owned in Ottawa, my then wife Mary and I endowed the 'Stephen and Anastasia Mysak Graduate Fellowship' in the Department of Atmospheric and Oceanic Sciences at McGill University [1]. My mother had died in 1978, but my father lived to see this tribute that November; he passed away peacefully at 100 in April 2007 (Chap. 1). Dad was always very interested in university life and often said "I would like to have been a professor." For the reception announcing the award, my former graduate student William Hsieh prepared a video of my dad talking about this fellowship, which is given annually to an MSc or PhD student doing research in oceanography, air-sea interaction or climate science. Since its inception, over 10 students have been recognized with this scholarship. It's been exciting and rewarding for me to meet these recipients and follow their careers.

When my wife Mary suddenly died in December 2011, my children and I decided to create a McGill prize in her honour. As a curious, mature student in McGill's Faculty (now School) of Religious Studies during the 1990s, Mary had a strong interest in eastern philosophies and religions. She felt that it was important to learn about the different people who came to our country from Asia. By matching the support of 75 donors (friends and relatives of Mary), we were able to establish in 2012 an endowment for the 'Mary Mysak Prize in Asian Religions', which is awarded annually to the graduating student with the highest standing in the Asian Religions program. Over the years, I have met brilliant recipients from Mauritius, India, Sri Lanka, Japan as well as Canada.

The next opportunity to establish other endowed awards came when I sold the family home in Montreal West in 2015. The net proceeds from this sale were shared equally between Paul, Claire, and me. I was delighted to be able to give the children this early inheritance, which they immediately invested. I decided to use my share of the sale to establish two undergraduate prizes in my name, reflecting my lifelong passions in science and music.

When I finished my mathematics degree at the U of A, I imagined that one day I would be able to return to my alma mater as a professor (Chap. 4). I loved teaching and explaining mathematics and science to others. However, as my career evolved into the fields of oceanography and climate science, I did not return to Alberta. After my retirement from McGill as a professor, I wanted to show my appreciation to the U of A, which gave me an excellent start as an academic. I decided to endow an undergraduate award in science for anyone studying oceanic, atmospheric or climate sciences. In an interview for *Faces of Philanthropy* published by the Faculty of Science at the U of A [2], I stated "my hope is that students can communicate their enthusiasm and understanding of science to others." Since science is increasingly under attack today in the public domain (especially in the USA), I believe this is an extremely important responsibility for both present-day and future scientists.

Each year a development officer from the Faculty of Science sets up a Zoom call with the recipient of the "Lawrence A Mysak Award in Science." I always look forward to these calls. I find them most fascinating and heart-warming as I learn about what has motivated these students to study environmental sciences and what they plan to do in the future.

Back at McGill in late 2015, I walked into the office of Kelly Rice unannounced. Kelly was then Director of Development for the Schulich School of Music at McGill. "I want to endow a prize for flute students," I said. "Wow," he exclaimed. "We don't have *any* prizes nor awards for wind students. This would be a first. However, would you consider expanding the terms of the prize to include any wind student?" he asked. "Fine with me," I replied.

Happily, for the first few years, the annual "Lawrence A Mysak Prize for Woodwinds" went to a flute student, and my wife Janet and I often went to recipients' graduation recitals. As a flutist myself, hearing these passionate performances brought tears to my eyes. In later years, the prize has been awarded to students of the clarinet and saxophone. A highlight for us each spring is the dean's reception for all the prize winners and donors to the music school. There we've had the pleasure of meeting the school's talented students and dedicated donors.

Since 2022, I have been supporting annually another music initiative at the Schulich School of Music, the "Mary Feher Community Engagement and Education Fund," which was established in 2021. Mary Feher and her late husband have been long-standing supporters of the McGill music school. Mary, a close friend of ours, established the above fund to support McGill music students visiting underserved Montreal communities to provide a variety of musical activities. Students receiving this award get valuable experience organizing these projects. Also, as these communities often have limited or no access to music or musical education, I feel that this is a very worthy cause.

33.1 Epilogue

Two decades on from establishing the first endowed award in 2006, I can honestly say that I've had much joy in giving with a "warm" hand rather than a "cold" one. I hope this story might inspire others to support their own chosen causes in the same way.

Assuming I haven't used all my pension fund and savings before my death, I have left several legacy donations. Among others, these will be top-ups to the Mysak awards described above.

Acknowledgements I am grateful to Janet Boeckh for suggesting this topic as a final chapter in my Memoir and for her many helpful comments. The input and suggestions from Claudia Bierman, Mary Feher, Amal Mikdame, Claire Mysak and Paul Mysak are much appreciated.

References

1. Haralambous K (2006) Mysak honours parents with fellowship. McGill Reporter 39, issue 7, Montreal
2. Schmidt A (2024) Faces of philanthropy: Lawrence Mysak. Faculty of Science, University of Alberta, Edmonton

Appendix
Chronology of Main Achievements, Awards, and Honors (Lawrence A Mysak)

1976 Promoted to Full Professor, Department of Mathematics, University of British Columbia, Vancouver

1978 Publication of treatise *Waves in the Ocean* (Elsevier), with co-author PH LeBlond

1981 President's Prize for 1980, Canadian Meteorological and Oceanographic Society (CMOS), with co-recipient PH LeBlond

1984-86 Principal Investigator, Project MOIST (Meteorological and Oceanographic Influences on Sockeye Tracks)

1986 Elected Fellow, Royal Society of Canada (FRSC)

1986 Appointed Atmospheric Environment Service (AES)/NSERC Industrial Research Chair in Climate Science, McGill University, Montreal

1989 Appointed Canada Steamship Lines Professor, Department of Atmospheric and Oceanic Sciences, McGill University

1990 Appointed Founding Director, Centre for Climate and Global Change Research (C^2GCR), McGill University

1993 Elected President, Academy of Science, RSC

1995 Appointed Honorary Professor, University of Kyiv Mohyla Academy, Kyiv, Ukraine

1995 George Lemaitre Visiting Professor of Climate Dynamics, Catholic University of Louvain, Belgium.

L. Mysak, *Adventures in Climate Science, Ocean Waves, and the Flute*, Springer Biographies, https://doi.org/10.1007/978-3-032-19848-8

1995 Co-soloist with Michael Flanders in Vivaldi Flute Concerto in C major for Two Flutes with *I Medici di McGill Orchestra*

1996 Named Member, Order of Canada (CM)

1997 Visiting Professor, Japan Society for the Promotion of Science

1998 Patterson Distinguished Service Medal, AES, Canada

1998 Tully Medal for Oceanography, CMOS

1999 Inaugural Fellow, CMOS (FCMOS)

2000 Fellow, American Meteorological Society (FAMS)

2000 Fellow, American Geophysical Union (FAGU)

2000 Foreign Member, Academia Europaea (ForMAE)

2000 David Thomson Award for Excellence in Graduate Supervision and Teaching, McGill University

2002 Golden Jubilee Medal, Canadian Governors General

2005 Prix Michel-Jurdant (Environmental Sciences), Acfas (Association francophone pour le savoir), Québec

2006 Alfred Wegener Medal and Honorary Member, European Geosciences Union

2006 Prix Marie-Victorin (Natural Sciences), Gouvernement du Québec

2007 Elected President, International Association for the Physical Sciences of the Oceans, IUGG

2009 Distinguished Alumni Award, University of Alberta, Edmonton

2010 Inaugural McGill University Medal for Exceptional Academic Achievement

2010 Appointed Canadian Steamship Lines Emeritus Professor, McGill University

2010 (June 1–2) Lawrence Mysak Session on "Ocean and Climate Dynamics", held at the 44th Annual CMOS Congress, Ottawa, in honor of Mysak's retirement from McGill on May 31, 2010. Organizers: Professor William Hsieh and Bruno Tremblay

2013 Flute Soloist in Gluck's Dance of the Blessed Spirits with *I Medici di McGill Orchestra.* In Memorium: Mary Mysak

2013 Appointed Chair, Steacie Prize Selection Panel, Canada

2015 Inaugural Fellow (Honorary Membership), International Union of Geodesy and Geophysics (FIUGG)

Index

L. Mysak, *Adventures in Climate Science, Ocean Waves, and the Flute*, Springer Biographies, https://doi.org/10.1007/978-3-032-19848-8

F

G

H

I

J